PUFFIN BOOKS

MISSION MOON

S.K. Das is Honorary Advisor to ISRO. His book *Touching Lives: The Little Known Triumphs of the Indian Space Programme* (Penguin) won the Luigi Napolitano Book Award of the International Academy of Astronautics, and has been translated into many regional languages.

MISSION MOON

EXPLORING THE MOON WITH CHANDRAYAAN 1

S. K. DAS

Foreword by G. Madhavan Nair

PUFFIN BOOKS

An imprint of Penguin Random House

PUFFIN BOOKS

USA | Canada | UK | Ireland | Australia
New Zealand | India | South Africa | China | Singapore

Puffin Books is part of the Penguin Random House group of companies
whose addresses can be found at global.penguinrandomhouse.com

Published by Penguin Random House India Pvt. Ltd
4th Floor, Capital Tower 1, MG Road,
Gurugram 122 002, Haryana, India

Penguin Random House India

First published in Puffin by Penguin Books India 2010

10 9 8 7 6 5 4 3 2

ISBN 9780143331308

Typeset in Century Schoolbook by Eleven Arts, New Delhi

Printed at Repro India Limited

www.penguin.co.in

MIX
Paper from responsible sources
FSC www.fsc.org FSC® C047271

This is a legitimate digitally printed version of the book and therefore might not have certain extra finishing on the cover.

CONTENTS

Foreword vi
Introduction viii

PART ONE

KNOWING OUR MOON

Exploring the Moon 1
How Did the Moon Form? 12
The Mysteries of the Moon 15
Did the Moon Change? 22

PART TWO

INDIA'S MISSION TO THE MOON

Planning for the Mission 33
The Spacecraft 38
The Payloads 52
The Rocket 59
The Ground Station 70
The Journey 76
The First Results 89
The Legacy of Chandrayaan 1 92

PART THREE

THE FUTURE

What Next? 99

FOREWORD

The night sky with twinkling stars and a full Moon is a sight that delights everyone. Moon, chaand or Chandamama is a common object of fascination for young kids. Rarely do we go through childhood without listening to stories about the Moon told by mothers or grandmothers. In India, we have many mythological stories around the Moon. In fact, the Moon is worshipped not only in India but also by several communities across the world.

Be it the full moon, new moon or an eclipse of the Moon, human beings have always evinced keen interest in these celestial events. The advent of the space age has added new dimensions to our observations and understanding of the Moon. Human beings have landed on the Moon and brought back samples of lunar soil. Several countries have sent probes and satellites to the Moon in order to study it better.

India achieved a rare feat of sending a satellite to the Moon and inserting it in lunar orbit and also placing the Indian tricolour on lunar soil in 2008–09. The Chandrayaan 1 project generated tremendous interest not only among the scientific community, but also among children in our country. The fascinating pictures sent by Chandrayaan 1 have enthralled scientists and students alike. It is essential to publicize the achievements of the country in space technology so that it kindles interest in the youth to take up space science as a career.

Chandrayaan has made a significant breakthrough in the annals of scientific discovery of recent times by detecting the presence of water on the Moon. This, I am sure, will have far-reaching effect on man's quest towards undertaking planetary missions through establishment of lunar colonies and using the Moon as an immediate base. I am sure this will spur the interest of youngsters.

I am extremely happy to note that Shri S.K. Das's book on the Moon targeting children is being published by Penguin Books India. Shri Das has an impeccable ability of conveying complex things in a simple and lucid style laced with humour. He has brought out several interesting facets of the Moon in the book, like its evolution and exploration by various countries. In addition, he has described the Chandrayaan 1 mission, the spacecraft and its functioning in a simple way.

I am sure this book will be of immense interest to readers young and old. I compliment Shri Das for writing this book and Penguin Books India for publishing it.

October 2009 G. Madhavan Nair
Bangalore

INTRODUCTION

It was the summer of 1947. An independent India was about to be born. People all over the country were looking forward to freedom from foreign rule. In a few days, Jawaharlal Nehru, the first Prime Minister of the country, would address the nation at midnight and talk about the new nation's tryst with destiny.

Young Krishnaswamy Kasturirangan was lying on the roof of his grandfather's house in Ernakulam. He was fascinated by the beautiful sight of the Milky Way that was like a glittering necklace across the night sky. The stars twinkled at him, as if surprised to find a young boy peering at them so intently at that hour when velvety darkness enveloped everything around. He was particularly fascinated by the Moon—a friendly, bright disc that dazzled so brilliantly in the dark. Its magic and mystery fascinated young Kasturirangan no end.

He asked himself many questions. Why isn't the Moon visible every night? Does the Moon have day and night? What makes the Moon shine? What causes part of the Moon to be lit up? How far away is it? It looked as if it was close enough to touch. Would he ever touch it?

Every time young Kasturirangan looked at the Moon, his imagination soared. By the time he left Eranakulam and joined a high school in Mumbai, the urge to understand the secrets of the night sky had become a passion. He studied science in college and for his research, he chose atmospheric science as the subject.

He joined the Indian Space Research Organisation (ISRO) and rose to become its chairman in course of time. While at ISRO, he built satellites that looked at Earth and its oceans. He also built satellites that were used for television, telephones and studying the weather. He built one of the biggest satellite systems in the world, its reach extending from Europe to Australia to Japan.

When he became the chairman of ISRO, he succeeded in realizing the dream of his childhood. He planned a mission to the Moon in 1999. The mission was to put a satellite in an orbit around the Moon. By the time he left ISRO in 2003, he had given final shape to the moon mission.

* * *

In another corner of south India, but not very far from Ernakulam, another young, inquisitive mind had started fantasizing about the Moon. It was in the small village of Thirunanthikera and the child was Madhavan Nair. When he was being fed on the lap of his grandmother under the open sky, she often promised to get him the Moon if he was a good boy and polished off everything on his plate. The clear, full moon in that unpolluted atmosphere made the young Madhavan Nair wonder why there were so many strange shapes on its surface. It was as if he could spot on the surface whatever his mind imagined. One day it was a deer walking across; another day it was a rabbit jumping around. His love for the Moon and his desire to unravel the many unknowns about outer space kept growing.

Madhavan Nair studied engineering and joined ISRO. He built rockets that carried satellites into space. He built the Polar Satellite Launch Vehicle (PSLV), the most reliable rocket the world has seen. Rockets often fail, but PSLV is one rocket that has flown successfully for fifteen consecutive flights. PSLV is a good rocket but it did not have the muscle to climb to 36,000 km where the bigger satellites had to be put. For that, Madhavan Nair built the Geosynchronous Satellite Launch Vehicle (GSLV). India was the sixth country in the world to do so.

Madhavan Nair never forgot the dream of his childhood. When he took over from Dr Kasturirangan as the chairman of ISRO in 2003, he carried forward the mission to the Moon with equal fervour. It took him five years but Chandrayaan 1 finally became a reality in October 2008. It was a fitting tribute that Chandrayaan 1 was lifted to the Moon's orbit on a PSLV rocket, the one that Madhavan Nair had built.

* * *

India's first mission to the Moon is an amazing story. It is as much a story of India going to the Moon as it is of collecting scientific information about the Moon and Earth that will benefit the entire mankind. Above all else, it is a story of how two young dreamers from the backwaters of Kerala studied science, built satellites and rockets, and finally realized their dream of reaching the Moon.

PART ONE

KNOWING OUR MOON

EXPLORING THE MOON

The Moon is not like our planet Earth. It does not have seas, rivers or streams. No flowers bloom there. There is not even a tree. No dog has ever barked there. No dinosaur has roamed its surface. There is no wind to blow away its dust. People have never lived there and yet they have dreamt about it for centuries. They have created all kinds of myths about it.

THE MOON AS A GOD

In our mythology, the Moon is a god. He is called Soma, named after nectar that makes the drinker immortal. God Soma gallops across the feathery clouds in a lovely three-wheeled chariot drawn by black antelopes. In three of his four hands he carries a mace, a sacred drink, a lotus; and the fourth is raised as if to bless the world. He is the son of Varuna, the lord of the ocean.

Soma is a loving god in Indian mythology. He is pleasant to children as well as elders. For half of the month, 36,000 gods and goddesses feed on Soma so that they can remain immortal. After so many divine creatures feed on Soma, he becomes exhausted and the Sun has to nourish him with water from the ocean.

Ganesha and His Big Belly

There is a story in Indian mythology about Lord Ganesha and the Moon. Once Ganesha was invited to sup with the moon god. Ganesha ate heartily and stuffed his belly so much that when he got up to wash his hands, he lost his balance and fell. The moon god laughed loudly and ridiculed Ganesha. The elephant god felt so humiliated and angry that he gave a curse and made the moon god vanish. Lord Shiva asked Ganesha to change his curse and the moon god to apologize. Ganesha changed his curse to say that the Moon would not disappear at once but would diminish gradually. After being absent for a day, the Moon would regain its power slowly. The cycle would continue forever so that people would be careful and never make fun of Ganesha.

THE ETERNAL REVENGE OF RAHU AND KETU

This is a wonderful story that tells us how the gods churned the ocean to find the nectar that would make them immortal. The gods found the nectar and sat down to drink it. Rahu, a demon, disguised himself as a god and sat down between the Sun and the Moon to drink the nectar. But the Sun and the Moon found out about Rahu's presence in their midst and raised an alarm. Lord Vishnu hurled his discus at Rahu and cut off his head. But, by the time the discus found the head and cut it off, Rahu had already drunk the nectar. What happened next was rather strange. His immortal head was cut off from an immortal body. The immortal head was Rahu while the immortal body was Ketu. In our astrology, the north node of the Moon is called Rahu and the south node is called Ketu. Ever since, Rahu and Ketu are looking for a chance to wreak vengeance on the Sun and the Moon whenever they come near. It so happens that, at certain times of the year, they swallow the Sun or the Moon causing either a solar or lunar eclipse. In Sanskrit, this is known as *grahana* or seizing.

Aryabhata Breaks with the Past

About 500 years after the birth of Christ, a new school of thought offered another explanation for eclipses. This happened in Khagola, the famous astrological observatory at the University of Nalanda located near the present city of Patna. When the bells in the university tolled at noon on 21 March 499 AD, the priests led a twenty-three-year-old astronomer to a platform. The astronomer sprinkled holy water on the parchment and pen lying on the platform. He chanted holy verses, gazed at the Sun overhead and prostrated himself in obeisance. Then he sat at the desk and wrote out his treatise. The astronomer was Aryabhata. He was born in Kerala and had come to complete his studies at the University of Nalanda. The treatise he wrote was considered a masterpiece and he was made the head of the university. Aryabhata said that the Earth was round and it rotated on its own axis, creating day and night. He also said that the Moon was dark and shone only because of the sunlight that fell on it. Solar and lunar eclipses, he said, occurred not because Rahu and Ketu swallowed the Sun and the Moon, but because of shadows cast by the Earth and the Moon. It was to recognize his great contribution to astronomy that India's first satellite was called Aryabhata.

WRITERS' FANTASY

Writers are very fond of the Moon. Beautiful poems have been written about it. In these poems, poets write about the Moon, especially the full moon, as a thing of great beauty. In one of his poems, Kalidasa, the famous Sanskrit poet, asks the king to give him a hearty meal that should have curds made from buffalo milk that looks like moonlight during the monsoon. Even for science fiction writers, from the Greek writer Lucien of the second century AD to Jules Verne in the nineteenth century, the Moon has been the favourite destination of their exploring heroes.

GETTING TO KNOW THE MOON

For mankind, the Moon has been the most fascinating object in the night sky. But we did not know what the Moon was really like till the telescope was invented about 400 years back.

Galileo sees oceans which are not there

Galileo, an Italian astronomer, was the first person to study the Moon through a telescope. When he turned his small telescope on the Moon in the year 1609, he was surprised to find mountains and valleys there. But he saw some dark areas, too. He thought these were oceans and seas with lots of water. Even other astronomers of that time thought these were large water bodies. So they gave them names like Sea of Tranquillity, Ocean of Storms and Sea of Serenity. But now we know these were not oceans or seas, or even water bodies.

The Moon gets a companion

The space age began on 4 October 1957 when Sputnik 1, the world's first satellite, was launched. Sputnik 1 started circling Earth. The Moon, which had circled Earth all alone for billions of years, got a tiny companion. The space age marked the beginning of a period that saw satellites going to the Moon, taking photographs and sending them back to Earth. These gave us answers to some of the basic questions about the Moon.

Yuri Gagarin of the Soviet Union was the first man to travel in space on 21 April 1961. He completed one full orbit in 104 minutes and came back to Earth in perfect shape. Looking at Earth, he was able to distinguish islands and rivers, shores and seas. He could even tell ploughed land from meadows. The Sun was ten times brighter than on Earth. The stars were bright and distinct, with much greater contrast than when seen from Earth. But Gagarin did not see the Moon.

THE RUSSIANS STRIKE THE FIRST GOAL

1959 was the year of the Moon. On 2 January 1959, just about a year after Sputnik 1 flew in space, Luna 1 headed for the Moon. Luna 1 weighed a little over 360 kg. It had several firsts to its credit. It was the first spacecraft to fly by the Moon. Then, it started going around the Sun. It was also the first spacecraft to orbit around the Sun.

The first man-made object on the Moon

On 14 September 1959 the first man-made object reached the Moon. It was Luna 2, an unmanned Soviet probe. On 4 October 1959, the second anniversary of the launch of Sputnik 1, Luna 3 went around the Moon and took photographs of the other side of the Moon never before seen by human beings.

The first lander and the first orbiter

Luna 9, another Soviet probe, was the first spacecraft to make a soft landing on the Moon on 5 February 1966 and send photographs back to Earth. The very next month, Luna 10 became the first lunar orbiter.

The Adventures of Luna 2

Luna 2 hit the Moon two minutes later than estimated. It broke into pieces as it hit the Moon, but it released on the Moon's surface two spheres covered with Lenin's pictures and the hammer-and-sickle emblem of the Communist Party. Luna 2 confirmed that there was neither any magnetic field within 55 km of the Moon's surface nor any trace of radiation belts.

The first robot

Luna 16 was the first robot to land on the Moon. It carried a drill and two electric motors for rotating the drill. When Luna 16 landed on the Moon, its robotic arm carried out its task by drilling the surface to a depth of 35 cm. Luna 16 brought back samples of the Moon's core to Earth.

The first rover

Lunokhod 1 was the first rover to land on the Moon. It was built like a cylinder with four wheels underneath. During its journey on the surface of the Moon, the brake of the wheels of Lunokhod 1 got stuck but it managed to travel a total of about 11 km, took 20,000 pictures, and mapped 80,000 square metres of the lunar surface. It was the most successful Soviet lunar mission.

A Bathtub?

The body of Lunokhod 1 was shaped like a bathtub which contained the computer, cameras, scientific instruments and the environmental control system. Though it was designed to work only for ninety days, it worked for eleven months. It was even spotted by the American astronauts who were flying overhead on Apollo 15 when sunlight glinted from the body of Lunokhod 1 roving on the Moon.

THE AMERICANS STRIKE BACK

At the time the Luna missions (twenty-four of them) were taken up by the Soviet Union, United States responded by carrying out the Ranger, Surveyor and Apollo missions. Apollo 8 was the first to carry human passengers to orbit around the Moon. Apollo 10 was the first to dock in the lunar orbit. Apollo 11 was the historic flight that carried humans to the Moon.

Apollo 11 took off on 16 July 1969 with three astronauts: Neil Armstrong, Edwin Aldrin and Michael Collins. The commander for the mission was Neil Armstrong. Michael Collins was the pilot of the Command Module while Buzz Aldrin was the pilot of the Lunar Module.

Giant leap for mankind

Three days after lift-off, Apollo 11 started orbiting the Moon at a distance of 100 km from its surface. Next day, Armstrong and Aldrin got into the Lunar Module (Eagle) and separated, leaving Collins in the Command Module. After about two hours, the Eagle touched down gently on the Moon, with seconds of fuel to spare. As Armstrong set foot on the lunar surface, his first words were, 'One small step for man. One giant leap for mankind.'

In Peace for All Mankind

On the Moon, Armstrong and Aldrin set up a television camera, planted an American flag, and collected soil and rock samples. They also unveiled a plaque which read, 'Here men from the planet Earth first set foot upon the Moon, July 1969 AD. We came in peace for all mankind.' The plaque had the signatures of all the three astronauts and American President, Richard Nixon. It also carried messages of most of the world's heads of state, and medals in memory of Yuri Gagarin and Vladimir Komarov.

The space buggy

After Apollo 11, six more manned Apollo missions were taken up, out of which five were successful. For Apollo

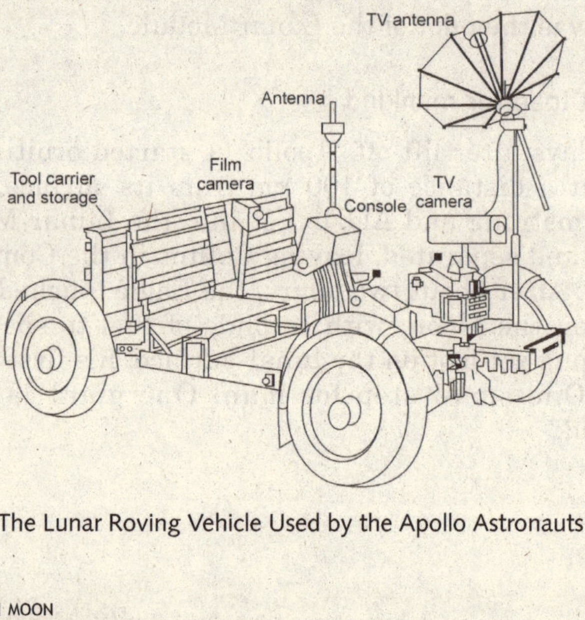

The Lunar Roving Vehicle Used by the Apollo Astronauts

missions 15, 16 and 17, the astronauts used a roving vehicle on the Moon which was a battery-powered space buggy. They used it to explore the landing sites and to travel greater distances than the astronauts on the earlier missions.

The last men on the Moon

By the time the last Apollo mission ended, twelve men had walked on the Moon over a period of four years. Between them, they had spent almost 300 hours on the Moon. The Apollo missions had achieved many things including collection of rock and sediment samples from the Moon. Analyses of the samples continue to give us new insights to the forces that shaped the early solar system, the Moon and more importantly, Earth.

THE MOON GETS BUSY AGAIN

There was a lull of almost ten years before the Moon got busy again in 1990. Galileo, a mission launched by NASA, the US space agency, took pictures of some of the regions of the Moon which had not been explored before.

In search of ice

The Clementine mission, launched in 1994, jointly sponsored by US and Europe, and the Lunar Prospector mission launched by NASA in 1998, had a common objective: to look for water/ice on the Moon.

Looking for hydrogen

Lunar Prospector also looked for water/ice on the Moon. For the purpose, it carried an instrument designed to measure neutrons. By looking at the energy of neutrons coming from the polar region, the Lunar Prospector team found excess of hydrogen at both the poles. The team estimated that, if it was in the form of ice, a total mass of 3,000 billion tonnes of ice could exist at the poles of the Moon.

A Clever Trick

The Clementine spacecraft did not carry instruments to specifically look for ice. But the mission team did a very clever thing. They improvized an experiment to look for ice. They beamed radio waves into the polar darkness and listened for echoes that would come by contact with ice. Such echoes came from the dark areas near the south pole of the Moon. Based on this, the Clementine mission found there was a possibility of finding large deposits of ice in the south polar regions of the Moon that were permanently shadowed.

Water is necessary for habitation on the Moon and therefore, efforts are being made to find out if there is water on the Moon. Water was not found in the rocks and dust brought back by the Apollo missions. The Lunar Prospector found evidence of hydrogen-bearing material in the polar regions of the Moon. Eleven years after the Lunar Prospector mission, the Chandrayaan mission finally discovered water on the lunar surface.

DON'T WE KNOW ENOUGH ABOUT THE MOON ALREADY?

So many probes and satellites have explored the Moon successfully. Men have walked on the Moon and brought back soil and rock samples. Then why did India take up a new mission to the Moon? Don't we already know a lot about the Moon, thanks to all these missions that have explored it in the past?

Mysteries of the Moon

True, we know more about many aspects of the Moon than we know about any world beyond our own. And yet, mysteries remain. People think that the Moon has already told us all the important things it has to tell. Nothing can be further than the truth. There are many exciting questions about the Moon for which we do not have definite answers.

Some of these questions are:

- Is it true that the Moon formed because of a giant collision of a big body with Earth?
- Once formed, how did the Moon gradually change over a period of time?
- In what quantities are minerals present on the Moon and where?
- Can helium-3 on the Moon be extracted and used as a clean and efficient fuel?
- Is there water/ice on the Moon?
- What kind of dangerous radiations surround the Moon and how strong are they?

India's mission to the Moon

India's mission to the Moon intended to find answers to these questions. The data that we have about the Moon is limited. It does not cover the entire Moon. In some cases, the data is not very clear. As a result, there are gaps in our understanding. It is only by going to the Moon to carry out new scientific explorations that we can hope to close the gaps in our understanding.

HOW DID THE MOON FORM?

How the Moon was born has been a puzzle for hundreds of years. It has continued to baffle astronomers and scientists for a very long time. Based on the samples of the moon rocks that the various missions to the Moon brought back, scientists believe that some of these rocks formed about 4,530 million years ago. This was about 30–50 million years after the origin of the solar system.

CRAZY IDEAS?

The mother and child idea

Scientists have suggested several theories about how the Moon formed. One of the earliest was the fission idea. It suggested that the primitive Earth was spinning so fast that a part of it broke off. This became the Moon. This is like a mother and child idea in which the child Moon is born out of the mother Earth. There are two problems with this idea. First, it would require the primitive Earth to be spinning incredibly fast: once every 2.5 hours. Second, one has to explain why Earth is not spinning that fast now.

The lover idea

It has also been suggested that the Moon was formed somewhere else and it was captured subsequently by

Earth's gravity. This is termed the lover idea in which a person captures a stranger by chance and gets tied to the stranger. This idea is a bit too improbable. If the Moon had a close encounter with Earth, it would have either meant a collision with Earth or the possibility of the Moon being thrown into a different orbit around the Sun.

The brother and sister idea

Yet another idea suggests that Earth and the Moon were formed together at the same time and place. This means the Earth and the Moon were born out of the same parent cloud. This, of course, would explain why we call the Moon Chandamama (Uncle Moon), but it does not explain things like Earth's rotation rate and how the material that formed the Moon got into orbit around Earth and stayed there instead of falling down to Earth.

All these ideas about the origin of the Moon have one defect or the other. The problem is so annoying that some scientists thought that all these scientific reasoning had perhaps proved that the Moon itself does not exist. But it does exist.

A BIG BLOW TO THE GROWING EARTH

The giant impact idea

The most probable theory about the formation of the Moon suggests it formed as a result of a big collision. A projectile hit Earth at a time when it had grown to 90 per cent of its present size and the iron core had already formed. It blasted off large quantities of heated material into orbit around Earth. The Moon formed from the debris.

This explains several things like the way the Earth spins; why Earth has a larger metallic core than the Moon; and why the Moon is depleted in iron. Moreover, modern theories about how planets were assembled from smaller bodies, which were in turn assembled from still smaller ones, predict that when Earth was almost formed, there would have been a body nearby with a mass about one-tenth that of Earth. So the giant impact idea is not really as strange as it seems.

The violence of the impact

Just consider how violent the impact would have been! It would have been nothing short of a catastrophe. It is difficult to imagine the energy involved. Much more than a trillion trillion tonne of material would have become vapour and melted. In the cloud around Earth, the temperatures would have exceeded 10,000 degree Centigrade.

The problem is that the giant impact idea is probable but not validated. It is a theory in need of a watertight proof. The proof can come from the photo-geological, mineral and chemical maps of the Moon's surface. These maps would allow scientists to distinguish different geological units that would give them some clues about the formation of the Moon by giant impact.

THE MYSTERIES OF THE MOON

If we look at the pictures of astronauts on the Moon, we see them picking up rocks from the surface. The astronauts were actually doing much more than collecting rocks. They were travellers in time. The rocks they brought back gave us clues to how the Earth and the Moon formed, the nature and timing of early melting, what happened when other bodies from space bombarded the Moon, and even the history of the Sun. These are very important clues to our understanding of not only the Moon but also of our Earth.

WHY CAN'T WE GET THIS INFORMATION FROM THE EARTH?

It is difficult to find this information on Earth because our planet has been so active that it has lost much of its historical record. So many mountains have been built, so many volcanoes have erupted, and there has been so much weathering and erosion in the billions of years of Earth's history that the clues have been lost. In our early history, four billion years ago, so much rain fell and so many tectonic planes collided that much of our planet's history got erased.

The Moon was luckier

The Moon was geologically active in its early days, but the changes in its physical structure were not vigorous and all records of early events were not lost. After the early volcanism ceased, there has been no volcanic activity for a long time. The footprints of the first men on the Moon still remain on its surface and shall perhaps be there for a million years.

The record keeper

The secrets of the Moon are recorded in its craters, plains and rocks. Since it is not always possible to find out what happened to the Earth millions of years back, the craters, plains and rocks on the Moon hold many important clues about the formation of the Earth and the Moon. The Moon holds many clues to our origin that can be revealed only by a proper study of its surface materials and its interior.

THE SURFACE OF THE MOON

If you look at the Moon long enough, you will see all the major features of the Moon's surface. One feature that stands out is the bright highlands. This is called the lunar terrae. Terrae is the Latin word for land. You will also see the darker plains. They are called the lunar maria. These names, terrae and maria, were given by Galileo about 400 years ago.

Galileo's small telescope

When Galileo looked at the Moon with his small telescope, he saw mountain ranges which he thought were the highlands and some plain, dark, nearly circular, low-lying basins which he thought were the seas. We know now that these are not seas or water bodies but we still use the term maria to describe them.

The pits in the beautiful face

What look like pits on the Moon's face are craters. The highlands have many craters that range in size from 1 metre to more than 1,000 km. All these craters would have been formed when objects bombarded the Moon. These objects are called projectiles.

The turbulent impact

The impact of projectiles hitting the Moon would have been very turbulent. You can imagine the impact, as these projectiles move at a speed of 20 km per second or 70,000 km per hour when they hit a surface. The projectiles themselves would have been so smashed that almost all of them would have become vapour, never to be seen again. The impact on the Moon's surface would have been equally great. Materials at the impacted site would have created piles of rock (called ejecta) around the circular hole and bright streaks of materials (called rays) would have been thrown over great distances.

Anatomy of a crater

Most craters on the Moon are small. They have a simple shape like a bowl. The bigger ones have more complex shapes with well-formed rims that rise above the ground.

When we count the craters in different areas of the Moon, we find that the bright highland areas have more craters than the dark maria. On the basis of this, scientists think that the highlands were formed first and the maria much later.

The mysteries of the dark seas

Astronomers in the olden days believed that the craters were filled with water. Scientists now think they are full of solidified lava. What it means is that although the Moon does not have volcanoes now, it would have had some volcanic activity in the past. This activity might have taken place after the highlands formed and most of the craters came into being.

The powdery blanket

The surface of the Moon is grey and sandy with fine sediments. Impacts from projectiles over billions of years have ground the surface into powder. Since the Moon has no atmosphere, even a small meteorite striking it can be damaging. There are some stones lying on the surface, but they will not last long. After a few hundred million years, these stones will be reduced to powder by meteorite strikes.

The powdery blanket that covers the Moon is called the lunar regolith. It contains the record of impacts during the past several hundred million years. It could give us crucial information to understand the impact of bombardment on Earth during that time. That is also why it was an important part of the Chandrayaan mission to map the Moon's surface very thoroughly.

Reaping a Harvest from the Solar Wind

The lunar regolith also contains information about the history of the Sun. Stars emit huge amounts of particles called the solar wind. Solar wind is composed mostly of hydrogen, helium, neon, carbon and nitrogen. The solar wind particles strike the lunar surface and find their way into the mineral grains. These amounts increase with the passage of time. So we should be able to find out if the conditions inside the Sun have changed over time by studying these products of solar wind. The solar wind gases may be useful in yet another way. When we establish permanent colonies on the Moon, life-support systems necessary for the survival of people, animals and plants would require hydrogen and oxygen (for water), carbon and nitrogen. Oxygen is there in the lunar rocks, and the solar wind should provide the rest.

Moon rocks

There is a big difference between the rocks on the Moon and those on Earth. Most common rocks on Earth are sedimentary. They were deposited under the sea and eroded by water. There is no water and atmosphere on the Moon and therefore, the most common kind of rock there is igneous (melt rock) in nature. Samples of Moon's rocks have given geologists some important information about Earth itself. The rock samples helped geologists understand what Earth would have been like 4.5 billion years ago.

An ocean on the Moon?

Scientists guess that an ocean flowed across the Moon once upon a time. It is just an idea. It is yet to be proved. When the first samples of the moon rocks were brought back to Earth, scientists found that they were mostly of one mineral—feldspar. These rocks are called anorthosites. One way the anorthosites could have accumulated was by floatation in magma (molten rocks). The same process happens in lava lakes and in magma chambers beneath volcanoes on Earth. The early Moon, it is suggested, must have been covered by a global ocean of magma, now commonly referred to as lunar magma ocean.

On the basis of these samples, scientists think that when the Moon formed, it was covered by a layer of magma hundreds of kilometres thick. As the magma crystallized, minerals which were more dense than the magma sank while those which were less dense, like feldspar, floated. The dense minerals later melted again to produce the basalts that the maria is composed of. This is supposed to be the magma ocean. When it cooled and crystallized, it formed a crust of about 60 km.

The Moon gets its crust

Scientists guess that the magma ocean might have been at least 500 km deep. The first minerals to form in the magma ocean would have been iron and magnesium silicates, olivine and pyroxene. As they were heavier than the magma, they sank. After that, feldspar formed. Since it was lighter than the magma, it floated to the top. In course of time, it produced heaps of anorthosite which made the first lunar crust. The magma ocean phase might have ended about 4.4 billion years ago.

Who left orange glass beads on the Moon?

It may seem strange, but orange glass beads were found on the Moon. On Earth, when the amount of gas dissolved in magma becomes large, it escapes violently and causes an explosive eruption. The lava erupts in large fountains up to several hundred metres high and falls to the ground in small pieces.

This also happened on the Moon. A sample of this was brought back by an Apollo mission. The sample was called 'orange soil' because it consisted of many small orange glass beads. They are glass because they formed from quickly cooled lava droplets during explosive eruption of hot fragments of rock and lava

What happened inside the Moon?

While the crust was forming on the Moon, other types of magma that would form a part of the highland's crust began to form deep inside the Moon. Scientists do not yet know where inside the Moon and how deep. In any case, these magmas rose and made their way into the anorthosite crust, forming large and small rock bodies and even coming out to the surface. This period of the Moon's history seems to have ended 4 billion years ago.

The process that took place in the formation of the Moon is called differentiation. It is a process of chemical zonation caused by differences in the densities of minerals: heavy materials sink while less dense materials float. It is an important geological process that leads to the production of a wide variety of igneous rocks.

The big question is: did the Moon really have a magma ocean? The data from Chandrayaan may give an answer. It could say whether there was a partial melting on the Moon which formed a magma ocean and a solidified crust. If the Moon had a magma ocean, the same idea could be applied to Earth. Earlier, scientists thought that the planets assembled when they were cold, and heated up later. If the mission finds that the Moon had a magma ocean, it would make the scientists look in a new way at Earth's early history.

DID THE MOON CHANGE?

One question scientists are trying to answer is, did the Moon gradually change over a period of time? This is important because the Moon has been a witness to 4.5 billion years of history of the solar system. The way the Moon has preserved records of this history more clearly than any other body in our planetary system is fantastic. The Moon is useful because it is our most reliable source for deciphering the history of all other planets.

THE BIG BOMBARDMENT

Scientists think that while the magma ocean was flowing and the crust was building, projectiles bombarded the Moon. The result was that rocks were modified soon after they were formed and the crust was mixed to a depth of almost 20 km. The bombardment might have declined between 4.5 and 4 billion years ago and then increased again, creating all those large basins we see on the surface of the Moon now.

LAVA FLOWS AND FOUNTAINS OF FIRE

When the bombardment slowed down, the maria formed. Between 3.7 and 2.5 billion years ago, lava would have flowed across the surface of the Moon. This formed the maria and decorated the Moon's face. There would have been high fountains of fire that threw glowing droplets of molten basalt up to a few hundred kilometres.

What a pit!

After the volcanic activities in the maria stopped, there has hardly been any geologic activity on the Moon. The only activity was the impact from projectiles, and this has created some magnificent craters. An example is Tycho that is on the southern highlands and is 85 km across. Its walls rise 3 to 4 km above its floor. Its rim rises above the surrounding terrain and its floor sits below it. The smooth material on its floor consists of melted rock that flowed like lava. The crater also throws out bright rays of material across the lunar landscape and causes landslides.

It was on one of the landslide sites which Tycho's rays had caused that an Apollo spacecraft landed and took samples. These samples indicate that the crater Tycho formed about 110 million years ago. It is fantastic how geologists sitting here on Earth can tell the age of a crater located on another planetary body!

ANATOMY OF THE MOON'S INTERIOR

Based on the geologic history of the Moon and the various samples brought back by the moon missions, geologists think they have found out things about the Moon's interior which perhaps looks somewhat like in the picture in the next page. One significant thing about the Moon is that quakes that occur there are very small compared to Earth. These moonquakes occur about once a year. This points to the fact that the Moon is not geologically active now. This is a blessing in disguise because the Moon is now a perfect place for astronomers to set up their telescopes.

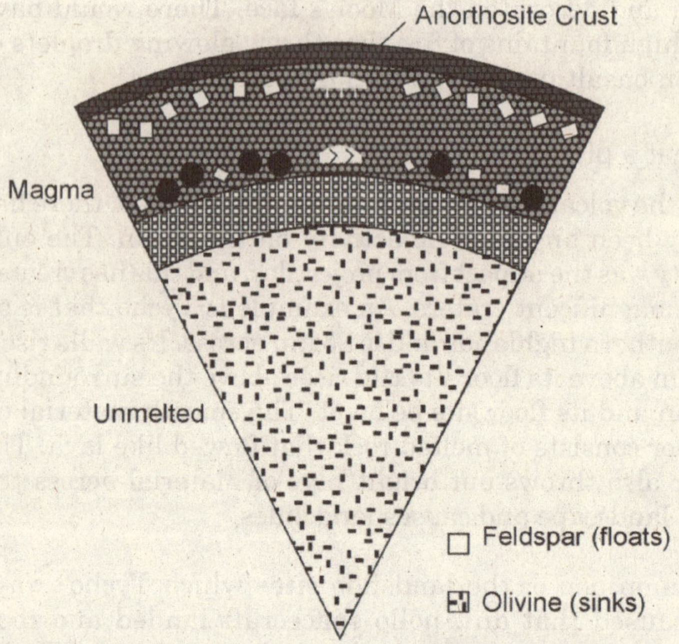

The Moon's Interior

IS IT LIKE THAT ON EARTH?

A comparison with Earth helps us to understand the nature of the Moon's interior. On Earth, we have a thin crust: 20–60 km over continents and 8–10 km over ocean basins, a thick metallic mantle (down to 2,900 km) and a large metallic iron core (2,900 km to the centre at 6,370 km). In the case of the Moon, the crust is thicker: from 70 km on the near side to 150 km on the far side. Below the crust is the mantle which is the largest part of the Moon. Beneath the mantle lies a small core made of metallic iron. We do not know about the exact size of the core. Scientists estimate it as ranging from about 100 km to 400 km.

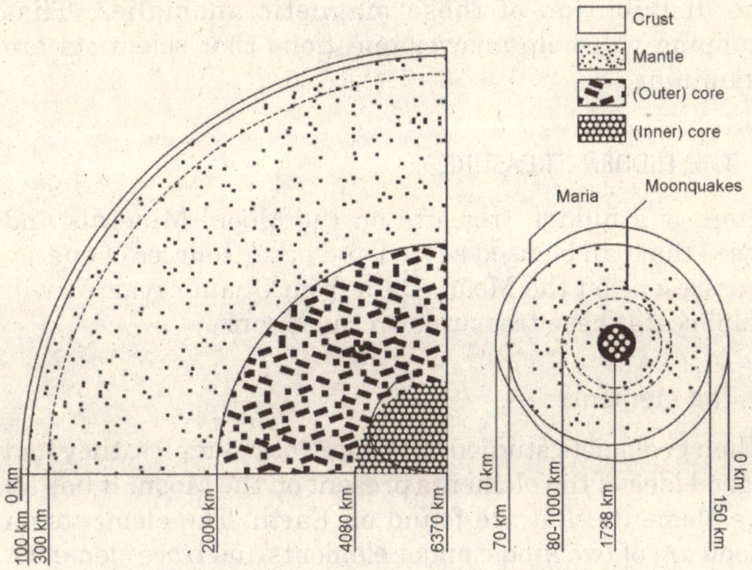

Earth and Moon Compared

WHERE DID THE MAGNETIC FIELDS GO?

The Indian mission seeks to find answers to two questions that are often asked about the Moon. Was there any magnetic field for the Moon? If so, what happened to the magnetic field? The answer has something to do with the smallness of the Moon's core. The Moon's core does not have much of a magnetic field like Earth's core has.

The magnetism of the Moon's core was not so weak always. The rocks that we find on the Moon's surface are magnetized. The older the rock, the stronger is its magnetism. This means that the Moon's magnetic field was stronger in the distant past. Why did it become weak?

We don't know the answer. There are small areas on the Moon that have stronger magnetic fields than the surrounding areas. The Indian mission proposes to map the distribution of these magnetic anomalies. That mapping will help answer questions that scientists are examining.

THE HIDDEN TREASURE

There is a hidden treasure on the Moon! Minerals and gases that can be used as fuel and other sources of energy are present on the Moon. Data from Chandrayaan 1 will help locate these treasures on the Moon.

The elements

When geologists studied the lunar rock samples, they had a good idea of the elements present on the Moon. It has all the elements that are found on Earth. The elements on Moon are of two kinds: major elements and trace elements. The major elements found both in the highlands and the maria are oxygen, silicon, aluminium, calcium, magensium, iron, sodium and titanium. We use iron, aluminium and titanium as construction materials, while silicon is used in the production of computer chips, solar cells, fibre-optic cables, mirrors and lenses.

Trace amounts of other important elements that are found are: sulphur, phosphorus, carbon, hydrogen, nitrogen, helium and neon. The trace elements are used to produce fatty acids, amino acids, vitamins and sugars (produced artificially) that are needed for life-support systems. The inert elements such as helium and neon can be used for compressed gas and other applications.

The wonderful world of helium-3

Helium-3 is the big find on the Moon. What exactly is helium? It is the gas we blow into a balloon. Helium has 2 protons and 2 neutrons in its nucleus. If you throw out one of the neutrons, you get helium-3. This happens once

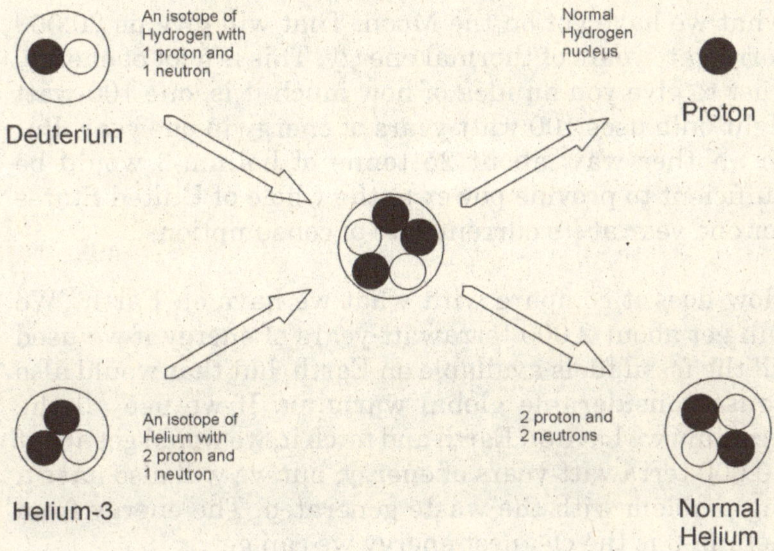

in a while in a powerful nuclear reactor like the Sun. The Sun produces helium by fusing hydrogen atoms together. But it so happens that about one in every 10,000 helium atoms comes out with one neutron. This is helium-3 and it can be used in reaction with deuterium to produce energy. One kg of helium-3 burned with 0.67 kg of deuterium gives us about 19 megawatt-years of energy.

The quantity of helium-3 available on Earth is very little. You can guess how little from the fact that the world-wide demand for helium-3 is 100 tonne a year but it is so difficult to get it that it sells for 3 billion dollars per tonne. But there is plenty of helium-3 on the Moon. Scientists have estimated that something like 11,00,000 tonne of helium-3 have been deposited by the solar wind in the Moon's regolith. The highest concentrations of helium-3 are in the maria of the Moon.

1 tonne of helium-3 gives us 10,000 megawatts of power/year. Suppose we use 10,00,000 tonne of helium-3 out of

what we have got on the Moon. That will give us 20,000 terrawatt-years of thermal energy. This is a lot of energy. Just to give you an idea of how much it is, one 100-watt light bulb uses 100 watt-years of energy in one year. Put in another way, about 25 tonne of helium-3 would be sufficient to provide power to the whole of United States for one year at its current rate of consumption.

How does it compare with what we have on Earth? We can get about 2,000 terrawatt-years of energy if we used all the fossil fuels available on Earth, but that would also cause considerable global warming. If we use all the uranium we have on Earth and torch it, we would get about 10,000 terrawatt-years of energy, but we will also have a big problem with the waste generated. The energy from helium-3 is the cleanest energy we can get.

RADIATION THAT KILLS AND MAIMS

The Moon has no atmosphere. Radiation from the Sun strikes the surface of the Moon with full force. Because of the high radiation environment, there is no life on the Moon.

One of the big things for the Indian mission is to study the dose of radiation on the Moon. There is a reason for this. Radiation is dangerous to human beings. When we build a colony on the Moon in the future, we need to have protection against radiation. The clothes of the astronauts have to be designed to protect them against radiation.

IS THERE WATER ON THE MOON?

It has been known for many years that the Moon has no atmosphere and hence, no surface water. But the meteorites that have been bombarding the Moon do contain water. They might have left water molecules on the surface of the Moon. Scientists now believe that these water molecules might have reached areas that are permanently shadowed from sunlight and got trapped there. Some of the deep craters in the polar region never get light from the Sun. They could possibly be the permanent traps of water molecules. It is in these craters that scientists expect to find water in frozen form.

The Chandrayaan mission has surprised everyone by finding water on the Moon, trapped a few millimetres under the lunar soil. Now that traces of water have been found, we do not have to transport water all the way from Earth to the Moon when we build a permanent human colony there. Water on the Moon would make living there easier, cheaper, and therefore, more likely.

GATEWAY TO THE SOLAR SYSTEM

From what we have discussed, it is easy to think of the Moon as a barren, airless rock in space. That would be a big mistake. Rather, we should think of it as mankind's stepping stone into the solar system. We should, one day, live there and from there visit destinations further into space. For that, we need to not only find resources to support human life but also to create infrastructure to travel to other places in the solar system.

PART TWO

INDIA'S MISSION TO THE MOON

PLANNING FOR THE MISSION

A DAY TO REMEMBER

It was 11 May 1999. India was celebrating Technology Day for the first time. On this day a year earlier, India had successfully tested a nuclear bomb.

At the function held at New Delhi, Dr K. Kasturirangan, chairman of ISRO, had been invited to deliver the Technology Day talk. The subject of his talk was India's Space Odyssey. He spoke on India's most successful rocket, the PSLV, and declared that the PSLV rocket could carry a mission to the Moon. He said, 'Working out the mission's objectives and payload could take time and if all goes well, it could be a reality by 2008.'

The time was just right. ISRO had, by 1999, established its strength in various areas of space technology. It had the capability to design, build and launch its own satellites with its successful PSLV rockets. It had developed and built science satellites and instrumentation. India was now poised to go to the Moon.

PUTTING HEADS TOGETHER

Dr Kasturirangan asked Dr Narendra Bhandari of the Physical Research Laboratory, Ahmedabad, to work on the scientific objectives for the mission to the Moon. Dr Bhandari was the right man. An eminent planetary scientist, he had spent his whole life studying the Moon. He had worked on the rock samples brought by the Apollo missions and the dust samples collected by Luna missions. In February 2000, Dr Bhandari and other scientists met at the Space Application Centre, Ahmedabad, to finalize the scientific objectives of the mission.

The idea of a mission to the Moon was discussed in several scientific bodies. The Indian Academy of Sciences endorsed it in 1999. It was also endorsed by the Astronautical Society of India in 2000. Based on the discussions in these two scientific bodies, Dr Kasturirangan appointed a Moon Mission Task Force with twenty experts to draw up a programme for the mission.

The Task Force did several things. It outlined the scientific objectives of the mission. It also drew up a list of scientific instruments to meet these objectives. It recommended that the existing ground infrastructure of ISRO needed to be strengthened for the Moon mission. It also suggested that for the first mission, the spacecraft should orbit around the Moon, and not land there. It even provided an estimate of the budget required for the mission.

In 2003, ISRO invited more than 100 scientists from all over the country to discuss the mission. The scientists were unanimous that the time had come for ISRO to take up the challenge. They also pointed out that an enterprise such as this will go a long way in helping develop an interest in space and space technology among the young minds of the country.

THE PRIME MINISTER ANNOUNCES THE MISSION

On Independence Day 2003, Prime Minister Atal Behari Vajpayee declared from the Red Fort in New Delhi, 'I am happy to announce that by 2008, India will send a mission to the Moon.' The mission was named Chandrayaan 1, making it clear that this would the first of many such missions.

What does Chandrayaan mean?

In Sanskrit, chandra means moon and yaan means vehicle. So Chandrayaan means lunar vehicle. The name was suggested by Atal Behari Vajpayee, Prime Minister of India (1996, 1998–2004).

THE WORK BEGINS

The formal approval of the Government of India came in November 2003. It had taken a little more than four years for the mission to come through since the day Dr Kasturirangan had proposed it in May 1999. But the wait was worth it. ISRO had used the time rather well. It had defined a detailed programme for the mission. It had spelt out the scientific objectives and proposed a list of scientific instruments to carry out these objectives.

THE MISSION IS DEFINED

The goals of Chandrayaan 1

- Expand scientific knowledge about the Moon.
- Upgrade India's technological capability.
- Provide challenging opportunities for planetary research to the younger generation of Indian scientists.

The scientific objectives of the Chandrayaan 1 mission

- Photograph the surface of the Moon in such a way that the length, breadth and height of the surface features of the Moon can be measured more accurately. This is called 3D imaging, and it leads to the preparation of a more accurate map. So an objective was to prepare a 3D atlas of the Moon's surface.
- Prepare an accurate map of the Moon showing how the various elements and minerals are distributed over its surface. The aim was to prepare chemical and mineral maps of the lunar surface for main rock-forming elements such as magnesium, aluminum, silicon, calcium, iron and titanium, and elements such as radon, uranium and thorium.
- Prepare a map of the Moon that shows different geological areas clearly.

These maps would help us understand matters relating to the Moon's formation by giant impact, the partial melting which might have formed a magma ocean and the solidified crust, the bombardment that might have formed the basins and the evolution of volcanic activities in the maria. As discussed in earlier chapters, there are important gaps in our understanding of the Moon in these areas. Chandrayaan 1 mission would help find answers to these fundamental questions.

PREPARING THE TOOL KIT

While selecting scientific instruments (or 'payloads' as scientists call them), ISRO was very particular that the instruments should be able to meet the objectives of the Chandrayaan 1 mission. These instruments included cameras for taking black and white 3D pictures or colour pictures, a LASER instrument to know the height of the Moon's surface features accurately, spectrometers to find out the availability of various radioactive and non-radioactive elements and various minerals, and a radar to detect the existence of ice/water in the polar regions of the Moon. These instruments not only rely on visible light, but also on infrared rays, X-rays and microwaves.

The International Conference on the Exploration and Utilization of the Moon met in Udaipur in November 2004. It was at this conference that ISRO displayed a model of Chandrayaan 1.

An Indian object to hit the Moon

Dr A.P.J. Abdul Kalam, then the President of India, attended the conference. Looking at the model of Chandrayaan 1, he suggested that the mission should not merely be an orbiting mission, but also a landing mission. That was how the idea of the Moon Impactor Probe (MIP) was born.

It was decided that the impactor, an instrument weighing about 29 kg, would detach itself from the orbiting Chandrayaan 1 spacecraft and rapidly descend towards the Moon's surface. During its flight to the lunar surface that would last for twenty minutes, it would take photographs of the Moon's surface, analyse the atmosphere and transmit the information to Earth via the orbiting Chandrayaan 1 spacecraft. Then it would crashland and while doing so, drop the Indian national flag on the surface of the Moon. This would make sure that our tricolour is placed on the Moon, making India one of the very few countries to have their national flags there.

The die had now been cast. ISRO was about to embark upon the first step that would eventually take India on a long, adventurous journey into space.

THE SPACECRAFT

Edgar Allan Poe, the American writer, talked about going to the Moon in a balloon. In Jules Verne's story, his three heroes travelled to the Moon in a shell fired by a cannon that was 274 metres long. The stories were, no doubt, very exciting to read but they were not exactly based on scientific principles.

HOW THE PLANETS MOVE

The first scientific principle of planetary motion was formulated by Johannes Kepler (1571–1630). He was a German astronomer. He suggested that the way planets moved around the Sun followed three laws. These are called Kepler's laws of planetary motion.

Kepler's Laws

- The orbit of every planet is an ellipse with the Sun at one focus of the ellipse.
- A line joining a planet and the Sun sweeps out equal areas during equal intervals of time.
- The square of the orbital period of a planet is directly proportional to the cube of the semi-major axis of its orbit.

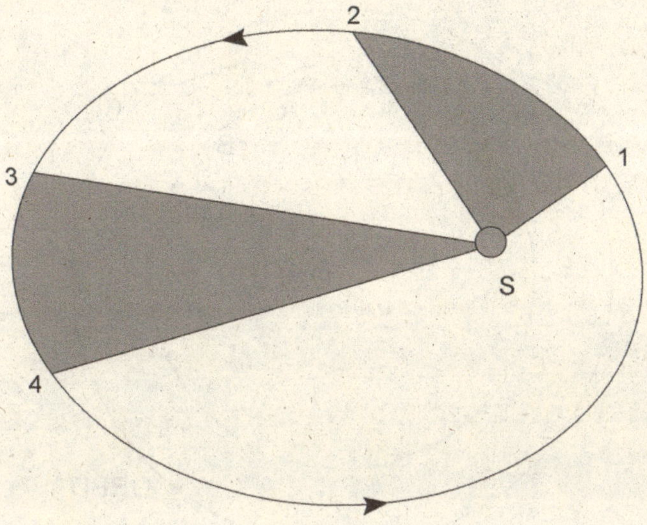

Kepler's Laws

The diagram above shows that as a planet goes around the Sun in an elliptical orbit, the time taken to go from 1 to 2 is the same as from 3 to 4 and the shaded areas are equal. Kepler helped us understand the motion of the planets. He told us how the planets moved, but he did not tell us why.

WHY DO THE PLANETS MOVE?

It was Isaac Newton (1642–1727) who told us why the planets moved the way they did. It was because of universal gravitation. Newton said if the gravitational force between Earth and the Moon suddenly vanished (became zero), the Moon would fly off in a straight line. And, if the Moon suddenly stopped moving, it would fall down on Earth.

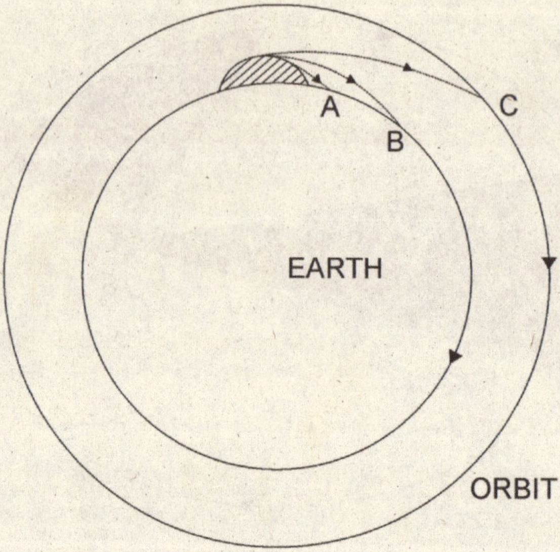

Newton's Idea of a Satellite of Earth

Newton also said if a body were thrown from the top of a high mountain with sufficient speed in a direction parallel to Earth's surface, it would go round Earth. In other words, it would become a satellite of Earth. The sufficient speed that Newton was talking about was 8 km per second. The diagram above tells us what Newton's idea of a satellite of Earth meant.

In the diagram, an object thrown from a high mountain in a direction parallel to Earth with some velocity, would fall on Earth at point A. With a higher velocity, it would fall at point B. With a velocity of 8 km per second, it would not fall on Earth but would become a satellite of Earth.

HOW DOES A SPACECRAFT ESCAPE EARTH'S GRAVITY?

If a spacecraft were launched from Earth's surface with a speed of 11 km per second, it would escape the gravitational force of Earth. This speed is called the surface escape velocity of Earth. For planetary exploration, spacecraft are first launched in Earth's orbit and then given higher speed to escape the gravitational force of Earth. The gravitational forces of the Sun, the Moon and other planets affect a spacecraft in its escape route. Therefore, the path of the spacecraft to its destination can be calculated only through mathematical calculations of orbit mechanics.

WHAT DEFINES AN ORBIT?

Three elements explain the orbit of satellites/spacecraft. These are the apogee, perigee and the inclination of the orbit. Apogee is the maximum distance from Earth. The perigee is the minimum distance from Earth. The inclination of the orbit is the angle between the orbital plane and the equatorial plane of Earth.

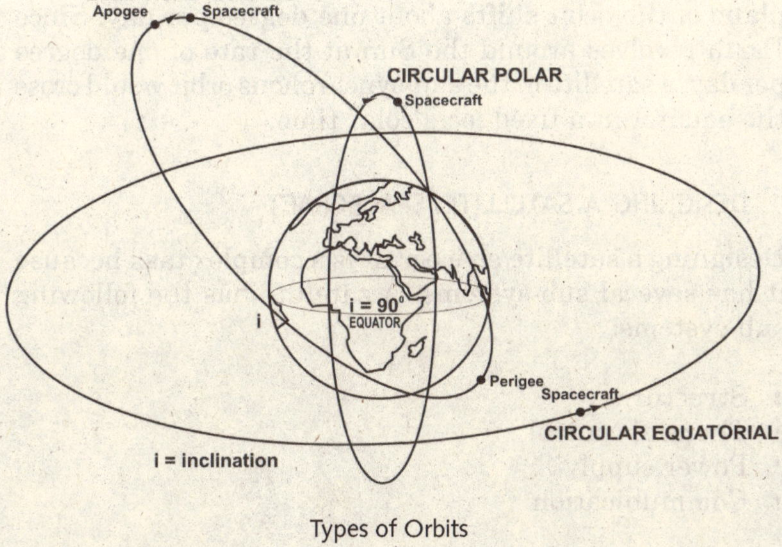

Types of Orbits

Polar orbit

If a satellite is in the polar orbit, it passes over the north or south poles. The satellite covers the entire Earth several times a day. Usually satellites that are designed to look at Earth and its oceans are put in the polar orbit.

Geosynchronous orbit

Sometimes a satellite is put in a circular orbit at a height of 36,000 km. Then the orbit is for a period of twenty-four hours which is equal to the time Earth takes to revolve around its axis. This orbit is called the geosynchronous orbit.

Geostationary orbit

If the plane of the orbit is in the equatorial plane of Earth, the orbit is called the geostationary orbit (GSO). That is because a satellite in this orbit looks as if it is stationary.

Sun-synchronous orbit

Another orbit is the sun-synchronous orbit. In this, the plane of the orbit shifts about one degree per day. Since Earth revolves around the Sun at the rate of one degree per day, a satellite in the sun-synchronous orbit would cross the equator at a fixed local solar time.

DESIGNING A SATELLITE/SPACECRAFT

Designing a satellite/spacecraft is a complex task because it has several sub-systems. A satellite has the following sub-systems:

- Structure
- Thermal control
- Power supply
- Communication

- Orientation
- Orbit control and orbit transfer

Structure

This houses everything that is kept inside the satellite. It has to be strong so that it can withstand the shocks and vibrations when the satellite is launched. It also needs protection from the harsh conditions in space like vacuum, extreme temperatures and radiation. The important thing is that the weight of the structure should be low. That is why materials that have low weight but high strength such as aluminium or magnesium alloys, titanium and fiberglass are used for making a structure.

Thermal control

The temperature of the various parts of the satellite is controlled with this sub-system. When in orbit, the outer surface of a satellite faces the Sun and it becomes very hot. The side facing away from the Sun becomes very cold. Also, the electronic instruments of the satellite become hot due to the electrical currents flowing within them. With the help of the thermal sub-system, the temperatures in different parts are controlled and kept within a narrow limit.

Power supply

The power supply sub-system provides electrical power to the various parts of the satellite. Power is provided by solar cells. The solar cells convert sunlight into electricity. The solar cells are kept on large panels on the outer surface of the satellite. These panels are made to face the Sun at all times. When the satellite is launched, these panels are folded on the sides, but after the launch they open out. Also, batteries are kept in the satellite to provide power during eclipses when the solar cells do not receive any sunlight.

Communication

This is the link between the orbiting satellite and the ground station located on Earth. It has three functions. It sends information about the functioning of the satellite to the ground station. This is called telemetry. It conveys commands from the ground station to the satellite to carry out certain functions. This is called telecommand. It also supports the ground station in tracking the satellite. The communication sub-system has a radio receiver, a radio transmitter and antennas to receive and transmit radio signals.

Orientation

For certain functions the satellite needs to be oriented in a specific direction. This process is called stabilization. A simple way of doing this is to spin the satellite along its axis. A satellite that is stabilized by spinning is called a spin-stabilized satellite. Another way of stabilizing a satellite is to keep three rotating wheels along the three mutually perpendicular axes of the satellite. This is called the three-axis stabilization technique. This technique gives better precision and is used in communication and earth observation satellites.

Orbit control and orbit transfer

Orbits of satellites are controlled very carefully. Sometimes it becomes necessary to make small changes in the orbit. To do that, small rockets called thrusters are fixed on the satellites. They are fired by giving commands from the ground station.

Orbit transfer is equally important. For example, a communication satellite reaches its geostationary orbit in two stages. First, the satellite is taken to an orbit with perigee at 200 km and apogee at 36,000 km. This is called the Geostationary Transfer Orbit (GTO). The satellite is then transferred to the geostationary orbit by firing a

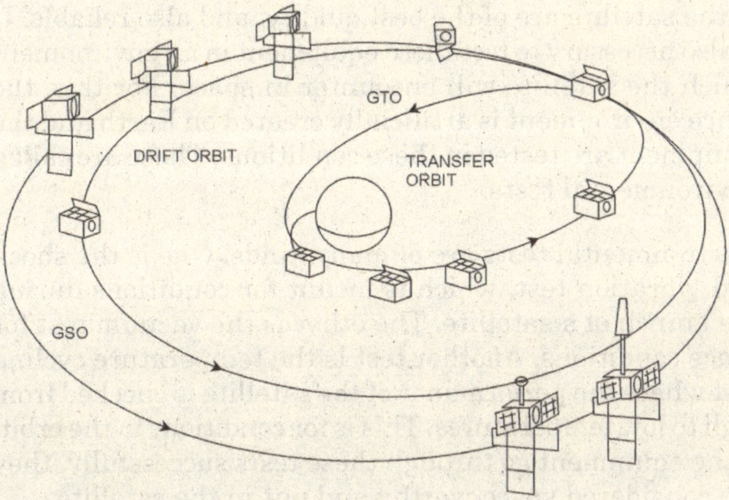

Transfer from Geostationary Transfer Orbit (GTO)
to Geostationary Orbit (GSO)

rocket engine which is fixed on the satellite. This is called the Liquid Apogee Motor (LAM) because it uses liquid propellants. The rocket engine is also used to give escape velocity to spacecraft meant for planetary exploration. The liquid propellants for the LAM and the thrusters are stored in the satellite/spacecraft itself.

HOW DO WE MAKE REPAIRS TO A SPACECRAFT IN SPACE?

Once a satellite/spacecraft is put in its orbit, it is not easy to make repairs to its parts. Of course, there have been cases when this has been done. This happened with a satellite called the Hubble Space Telescope. The astronauts on the Space Shuttle docked in with the Hubble, made repairs and replaced certain equipment. But this is a very expensive exercise.

Because we cannot make repairs to a satellite in space easily, it is necessary to make sure that all the equipment in the satellite are of the best quality, and also reliable. It is also necessary to test these equipment in an environment which the satellite will encounter in space. For this, the space environment is artificially created on Earth and the equipment are tested in these conditions. These are called environmental tests.

Environmental tests are of many kinds. One is the shock and vibration test, which is meant for conditions during the launch of a satellite. The other is the vacuum test for space conditions. Another test is the temperature cycling test where the performance of the satellite is checked from high to low temperatures. This is for conditions in the orbit. If the equipment go through these tests successfully, they are considered space-worthy and put in the satellite.

TYPES OF SATELLITES AND SPACECRAFT

Communication and weather satellites

A satellite in the geostationary orbit at a height of 36,000 km has an orbital period of twenty-four hours. This is exactly equal to the period of Earth's rotation. Such a satellite appears to be stationary and is always available for communication purposes. A satellite in this orbit can keep radio contact with all the ground stations located in one-third of Earth. It is used for television broadcast over large areas and point-to-point communication services.

The Pioneer

Arthur C. Clarke wrote an article in 1945 in which he said that three satellites placed in the geostationary orbit could provide telecommunication services to the whole world. At the time Clarke wrote his article, the idea of a satellite was still very new. It took a long time for scientists to translate his idea into action.

Sometimes weather satellites are also placed in the geostationary orbit to give information about weather. In that case, the satellite can continuously monitor weather conditions over one-third of Earth's area. ISRO has built a large number of satellites for communication and weather. It now has one of the biggest communication satellite systems in the world, its reach extending from Europe to Australia to Japan.

Remembering Kalpana Chawla

ISRO's weather satellite Kalpana 1 is located in the geostationary orbit. It was named in the memory of Kalpana Chawla, the astronaut who was killed in a space shuttle disaster in 2003.

Remote sensing satellites

These are satellites that survey resources of Earth like land and water from space. They also survey the oceans. The technique used for such survey is known as remote sensing. With this technique, measurements can be made from a distance without being in direct physical contact with the object to be measured. These remote sensing satellites are launched in the polar orbit at a height of 900 km so that they can cross the equator at a constant local solar time. ISRO has built and launched a large number of remote sensing satellites. Information from these satellites is used for surveying India's natural resources.

Scientific spacecraft

Spacecraft for scientific exploration of the Moon and planets of the solar system can be of three kinds: a fly-by mission, an orbiter or a lander. A fly-by spacecraft travels on a path which is at a distance from the planet and while flying by, it makes observations of the planet and takes photographs. An orbiting spacecraft goes in an orbit around the planet and the instruments in the spacecraft make observations and take photographs. A lander is first put in an orbit around the planet and a probe takes off and lands on the planet. A lander can also carry a rover that can travel over the surface of the planet and make observations.

THE SPACECRAFT FOR THE MOON

The Chandrayaan 1 mission was an orbiting spacecraft that housed and supported scientific instruments. It also carried a lander (MIP) that took off from the orbiting spacecraft and crashlanded on the surface of the Moon on 14 November 2008.

Structure

While designing the Chandrayaan 1 spacecraft, the prime requirement was that it should protect the scientific instruments in the harsh environment surrounding the Moon and provide the right conditions for the instruments to function. The spacecraft was shaped like a cuboid. Each side was approximately 1.5 metres. It weighed about 1,380 kg. More than half was the weight of the propellants. So, by the time the spacecraft reached its lunar orbit, its weight reduced to about 675 kg. It was a three-axis stabilized spacecraft.

The structure was built using composites and aluminium honeycomb material. It was strong enough to withstand

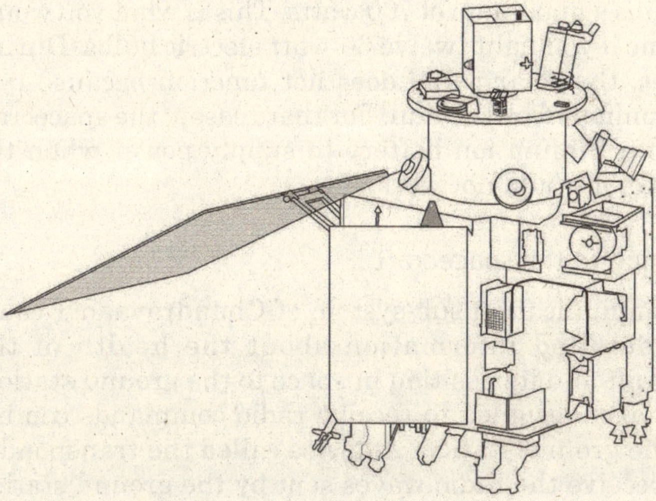

Chandrayaan 1 Spacecraft

shocks and vibrations when it was launched, and also function in the harsh conditions in space.

Keeping it cool

The thermal sub-system was designed to keep the temperature of the Chandrayaan 1 spacecraft at a comfortable level and allow the scientific instruments as well as the entire spacecraft to function efficiently and safely. For this purpose, different kinds of tapes, paints, multi-layer insulation blankets, optical solar reflectors, heat pipes, heaters and temperature controllers were used.

Borrowing power from the Sun

The power supply sub-system had a solar panel and battery. This provided the required power to the spacecraft during all phases of the mission. In the diagram of the spacecraft above, you see a solar panel projecting from one side of the spacecraft. This is a single solar panel that generates a maximum of 700 watts. This is what you would consume if you light twelve 60-watt electric bulbs. During eclipses, the solar panel does not function because it is not illuminated by the Sun. For that reason, the spacecraft carried a lithium ion battery to supply power when the solar panel would not work.

Talking to the spacecraft

The communication sub-system of Chandrayaan 1 could radio detailed information about the health of the spacecraft and its position in space to the ground station. It was also designed to receive radio commands coming from the ground station. A device called the transponder could receive the radio waves sent by the ground station antennas, modify them in a precise way and send them back. Through this, scientists on the ground knew where exactly Chandrayaan 1 was at a particular point of time. The spacecraft used a small, dish-shaped antenna called the dual gimballed antenna for transmitting data from the scientific instruments.

Keeping it up there

It was very important to know the direction in which the different faces of Chandrayaan 1 were pointing in space. It is important for any satellite, but more so for a spacecraft designed to orbit around the Moon because it helps in putting the spacecraft in the lunar orbit, maintaining that orbit or making changes in the orbit if necessary. For this purpose, sensors had been put in Chandrayaan 1. Some of these sensors took the direction of the Sun for reference, while some others took the direction of stars. These are called the Sun and Star Sensors. There were also small gyroscopes on the spacecraft. They provided the orientation reference for the spacecraft when in space.

The brain of the spacecraft

The attitude and orbit control sub-system is the brain of a spacecraft. It is called the Bus Management Unit (BMU). It ensures proper orientation of the spacecraft. It also helps scientists change its orbit during different phases of its flight. On Chandrayaan 1, flywheels called reaction wheels helped keep the spacecraft stable so that it would not shake while taking photographs of the Moon. Small rocket engines called thrusters helped in providing orientation to the spacecraft at certain times. To make Chandrayaan 1 escape from orbiting Earth and travel to the Moon, a Liquid Apogee Motor was used. Liquid propellants needed for the LAM as well as for the thrusters were stored onboard the spacecraft.

THE MEN WHO BUILT IT

The spacecraft was designed and built by engineers and scientists at ISRO Satellite Centre (ISAC). Located at Bangalore, it is ISRO's lead centre for satellite development. The electro-optic sensors required for the spacecraft were provided by the Laboratory for Electro-Optic Systems also located at Bangalore. Other centres of ISRO which contributed to the building of the spacecraft were: Vikram Sarabhai Space Centre, Liquid Propulsion Systems Centre, ISRO Inertial Systems Unit, Space Applications Centre and Physical Research Laboratory.

THE PAYLOADS

The type of payload that a satellite or spacecraft carries depends on its mission. For a communication satellite, the payload has communication equipment and antennas. For a scientific spacecraft, the payload has scientific instruments. How the instruments function would depend on the support given by the sub-systems of the spacecraft that we talked about in the last chapter.

THE PAYLOADS OF CHANDRAYAAN 1

There were eleven payloads (scientific instruments) through which the Chandrayaan 1 mission intended to achieve its objectives. Of these, five instruments were designed and developed in India. Three instruments were from the European Space Agency, one of which was developed jointly with India, and the other was done with Indian contribution. One was from Bulgaria. Two were from the United States.

MADE IN INDIA

Terrain Mapping Camera (TMC)

This was an electronic camera that took black and white pictures of the Moon's surface. In these pictures, objects on the Moon's surface as small as 5 metre (about 17 feet)

SLV-3 being transported to the launch pad (July 1980). On the right is Satish Dhawan, who was the Chairman of ISRO at that time. On the left you see A.P.J. Abdul Kalam, who was in charge of building the SLV-3 rocket and later became the President of India. G. Madhavan Nair, who was the project manager for SLV-3 under Kalam, was the Chairman of ISRO during Chandrayaan 1 mission.

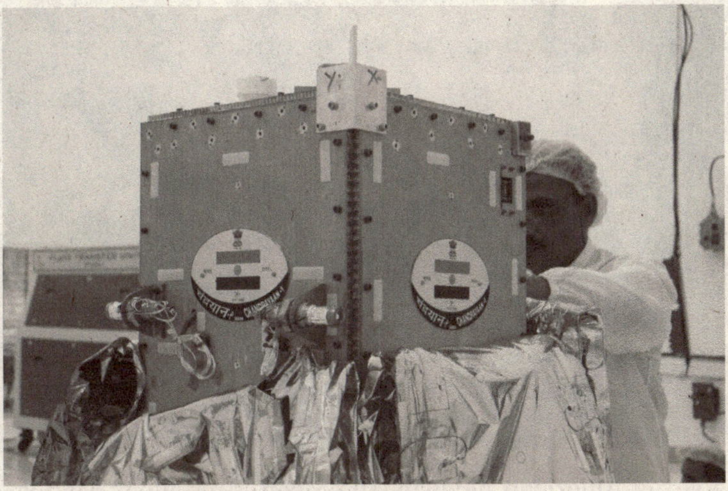

The Moon Impact Probe, one of the Indian payloads of the Chandrayaan 1 spacecraft, being prepared. It carried a video camera, a radio altimeter and a spectrometer.

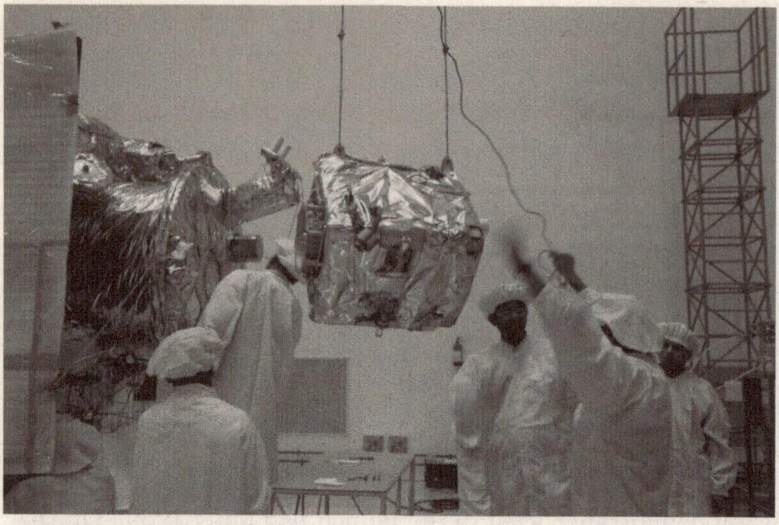

Payload preparation for the Moon Impact Probe. The probe was designed to separate from the orbiting Chandrayaan 1 spacecraft and hit the Moon's surface.

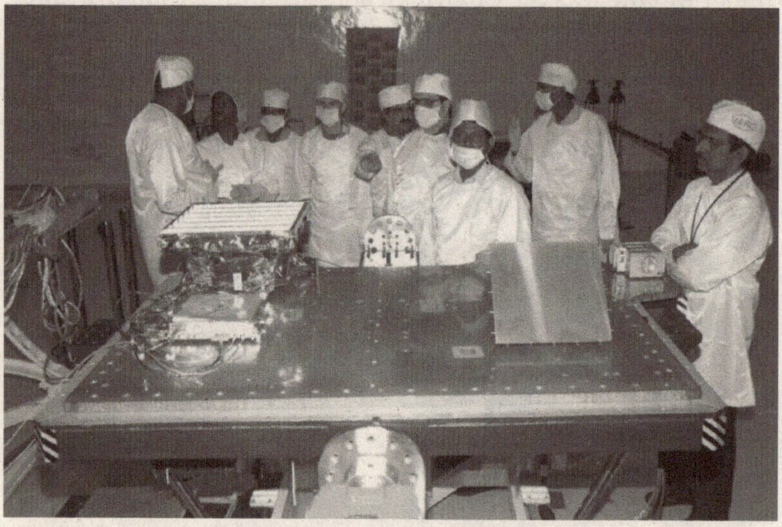

This is the American payload, the Moon Mineralogy Mapper, getting ready for being kept with the spacecraft. This instrument, jointly developed by Brown University and Jet Propulsion Laboratory of the US, discovered water on the Moon.

Final touches being given to Chandrayaan 1 spacecraft at the ISRO Satellite Centre, Bangalore.

PSLV-C11 stacked up to the Fourth Stage. At the top of the vehicle there is a bulbuous fairing. It has a diameter of 3.2 metre. It protects the satellite and is discarded after the launch vehicle clears the dense atmosphere.

PSLV-C11 at Vehicle Assembly Building. It is as tall as a fifteen-storey building (44 metre) and weighs 320 tonnes.

PSLV-C11 on its way to the launch pad.

PSLV-C11 on the Second Launch Pad.

PSLV-C11 going up in a blaze of glory.

The 18 metre Dish Antenna (IDSN-18) was established at Byalalu as part of the Indian Deep Space Network for Chandrayaan 1 mission operations and collection of data.

The first image taken by the Terrain Mapping Camera on 31 October 2008. It shows the northern coast of Australia.

Earth as viewed by Chandrayaan 1 Terrain Mapping Camera.

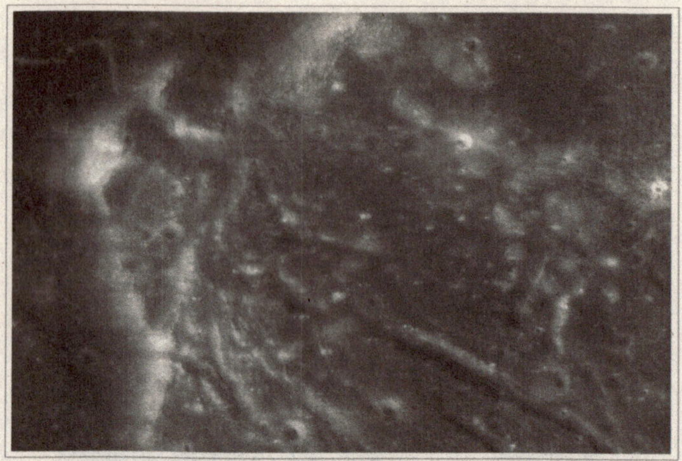

SCALE = 1:30000

Image of the Moon captured by the Terrain Mapping Camera.

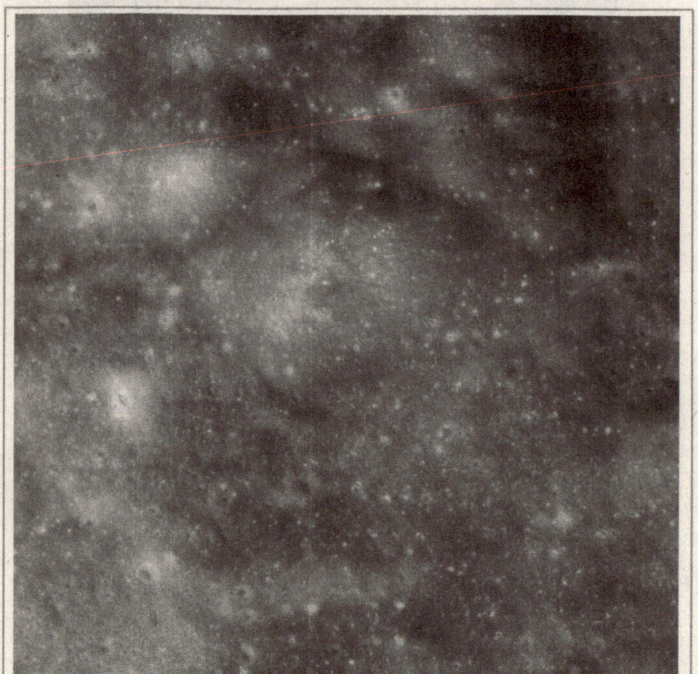

SCALE = 1:30000

Image of the lunar surface captured by the Terrain Mapping Camera. The height of the features shown in these images can be grasped if we view them through an anaglyph filter.

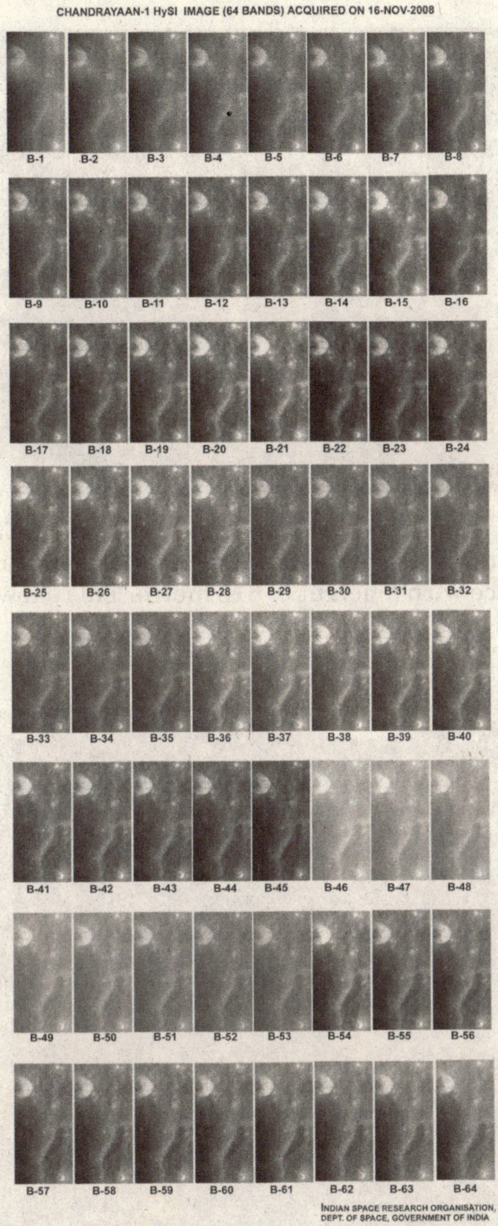

Pictures imaged by the Hyperspectral Imager of the Barrow crater on 16 November 2008. They show pictures as imaged by all the sixty-four bands of the Hyperspectral Imager.

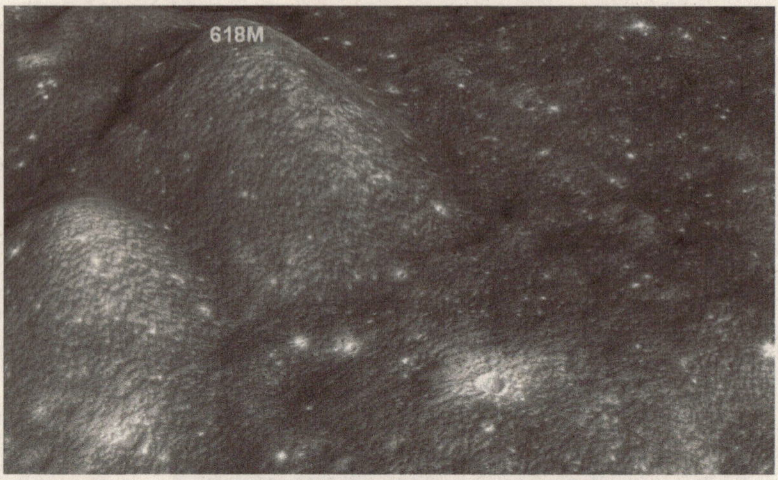

This picture was taken by the Terrain Mapping Camera on 23 November 2008. This is a 3D view of a mountain on the Moon's surface with a height of 618 metre. This shows an area of 10x10 km.

This picture was taken by the Terrain Mapping Camera on 23 November 2008. This is a 3D view of a crater of a depth of 400 metre. This covers an area of 10x10 km.

A digital elevation of the Moon's surface generated using Terrain Mapping Camera imagery.

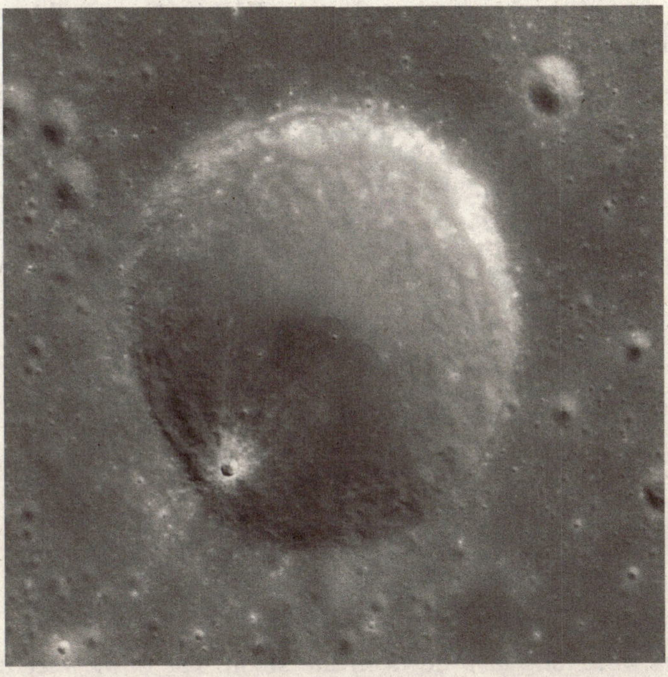

Finer details of a lunar crater captured by the Terrain Mapping Camera.

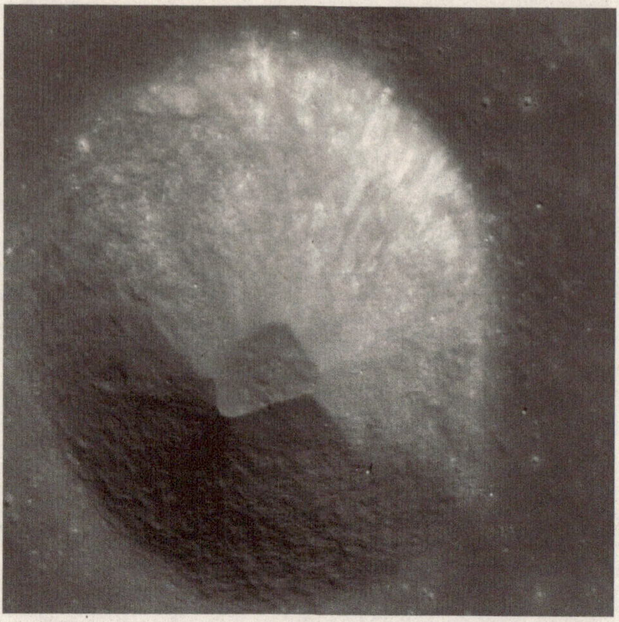

Far side of the Moon viewed at NADIR by Terrain Mapping Camera on 24 November 2008.

There were long lava flows on the Moon. The flows were up to 400 km long and have 30 metre high cliffs at their margins.

The polar region of the Moon captured by the Terrain Mapping Camera.

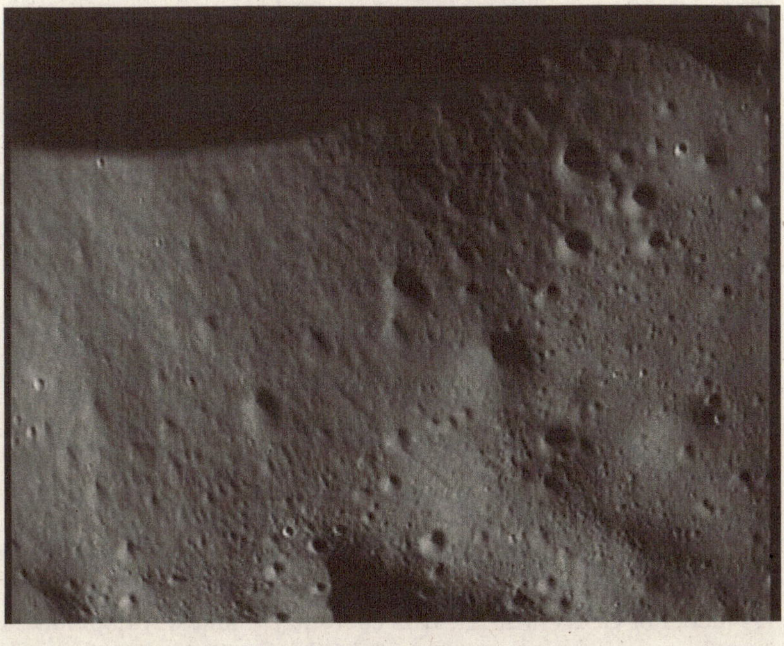

A close-up picture of the Moon's surface taken by the Moon Impact Probe before it crashlanded on the Moon.

A model of GSLV Mk-III. It is currently under development. It can launch a 4 tonne spacecraft in the Geostationary Transfer Orbit (GTO).

could be captured. The TMC was designed so that the height of the object could be determined by looking at the picture. In other words, the TMC took three-dimensional pictures of the Moon's surface. The aim of this stereo camera was to map the topography of both the near and far side of the Moon and prepare a 3D atlas. This would help us to understand how the Moon formed and the way it has changed with the passage of time. The atlas will also help in identifying areas on the Moon's surface which are of particular scientific interest and which require further investigation.

Hyperspectral Imager (HySI)

This camera took colour pictures of the Moon's surface by recording the visible and infrared light reflected from the Moon. Features that are about 80 metre (260 feet) can be identified in such pictures. The HySI split the light from the surface of the Moon into sixty-four thin slices that revealed a specific object or phenomenon. HySI, obtained data for mapping minerals on the surface of the Moon. The minerals present on the lunar surface will give us important clues to melting, differentiation, crystallization and volcanism. These are the lunar processes we had talked about in the earlier chapters.

Lunar Laser Ranging Instrument (LLRI)

This instrument sent out pulses of infrared LASER light towards the Moon's surface and sensed the reflected light. By doing this, it provided accurate information about the height of the lunar surface features. Knowing the height of the surface features of the Moon will help us in understanding the internal structure of the Moon. It will also help us in understanding how the Moon cooled during its history as well as how the surface features formed in the past.

LLRI could also tell the height of the spacecraft above the lunar surface. Chandrayaan 1 was expected to stay at a height of 100 km. The height keeps on changing because of gravity fluctuations on the Moon. If the height of a lunar orbiting spacecraft is not corrected from time to time, it will crash on the Moon. LLRI provided the ranging data to know the height difference between the spacecraft and the Moon's surface. It helped to determine the exact height and if the spacecraft exceeded the recommended limit, the spacecraft could be brought back to the normal height of 100 km.

High Energy X-ray Spectrometer (HEX)

This could sense the emission of X-rays having energy which the scientists consider as high. Such X-rays are emitted by heavy elements that are radioactive. By recording and analysing such X-rays, we can identify the presence of heavy elements like uranium and thorium. We can also identify the presence of water ice. HEX was looking at the polar regions of the Moon for ice deposits. It was also locating areas where uranium and thorium can be found in plenty.

Moon Impact Probe (MIP)

The MIP was designed to separate from the orbiting spacecraft and hit the Moon's surface. This device was put as a payload for three reasons. The first was to get experience in making an object hit a particular area on the Moon's surface at a particular time. The second was to test certain devices that would help the spacecraft in making a smooth landing on the Moon in future. The third was to find out more about the extremely thin atmosphere of the Moon.

The MIP carried three instruments. One was a video camera to take pictures of the Moon as it approached the surface before hitting it. The second instrument was a radar altimeter which sent out microwaves and sensed their reflection, and calculated its height from the surface of the Moon. The third instrument was a spectrometer which sensed the small quantity of gaseous atoms that surround the Moon.

On the whole, MIP was expected to validate some of the technologies required for soft landing on the Moon in the future.

PAYLOADS FROM OTHER COUNTRIES

There were six payloads from other countries in Chandrayaan 1. Three of them were from the European Space Agency.

Chandrayaan 1 X-ray Spectrometer (C1XS)

The name of the payload is pronounced as 'kicks'. It was a spectrometer like the Indian payload HEX measuring X-rays. But C1XS measured another type of X-rays that are emitted by certain elements on the surface of the Moon. X-rays coming from the Sun produce fluorescence in various elements present on the Moon's surface. The abundance of elements such as magnesium, aluminium, silicon, calcium, titanium and iron can be obtained with fluorescence X-rays. C1XS was mapping the Moon by using this technique. It mapped the highlands, the maria, impact basins and the large craters on the Moon. Data from C1XS will help us understand the existence and scale of pre-maria volcanic activity, to refine the estimates of bulk composition of the Moon and to improve the evolutionary models.

Smart Near Infrared Spectrometer (SIR-2)

Many minerals on the Moon reveal their existence by emitting infrared rays. Infrared rays lie next to visible light in the electromagnetic spectrum and are of a longer wavelength. SIR-2 could detect such rays and find out where and in what quantities these minerals are there on the surface of the Moon. The data it sent back can also help us to understand the way in which the different layers of the Moon's crust lie over one another. It will throw light on the way the materials on the Moon's surface have changed due to extremely hot and cold conditions there. On the whole, SIR-2 had to study the Moon's surface to survey mineral resources for landing sites and exploration.

Sub ke V Atom Reflecting Analyser (SARA)

The Moon does not have an atmosphere. The solar wind hits the lunar surface directly and as a result, atoms escape from the Moon's surface. SARA was designed to detect such atoms. Analysis of these atoms will help in understanding the relationship between the Sun and the Moon. It will also give us clues to the Moon's magnetic fields of the past. On the whole, it will help us to understand the surface composition of the Moon, the way in which the Moon's surface reacts to the solar wind, how materials are altered in space with time and the magnetic anomalies associated with the surface of the Moon.

Radiation Dose Monitor (RADOM)

This was the Bulgarian payload carried by Chandrayaan 1. Its function was to measure the radiation dose on the Moon. This will provide information on the quantity and distribution of radiation around the Moon. Radiation is dangerous to human beings. Study of radiation is important because when we build a habitat on the Moon, the radiation dose has to be assessed so that the habitat can be designed properly to protect the astronauts. Even at the time of manned missions, we should know the dose of radiation so that the astronauts' suits can be designed accordingly.

Mini Synthetic Aperture Radar (MiniSAR)

This instrument of the United States was flown in Chandrayaan 1. It was a special type of small radar. But its functions were almost the same as that of a radar with a big antenna. It had a special way to emit microwaves and sense their reflection. The idea of putting this instrument as a payload in the spacecraft was to detect water/ice in the permanently shadowed polar regions of the Moon. As we discussed earlier, the comets and meteorites that have been bombarding the Moon might have left water molecules. These molecules might be trapped in the deep craters of the polar region that never get light from the Sun.

Moon Mineralogy Mapper (M3)

This instrument too came from the United States. Sponsored by NASA, it was built by Brown University and Jet Propulsion Laboratory. The instrument was meant to identify and map minerals or polar resources such as water and other volatile components. This was the instrument that found water on the lunar surface. The water was found in molecular form, trapped a few millimetres under the lunar soil.

HOW WERE THESE PAYLOADS KEPT IN THE SPACECRAFT?

The diagram on page 58 shows how these eleven payloads were positioned in Chandrayaan 1.

All these payloads were selected to achieve the scientific objectives set for the Chandrayaan 1 mission. They made Chandrayaan 1 a comprehensive mission, fully equipped

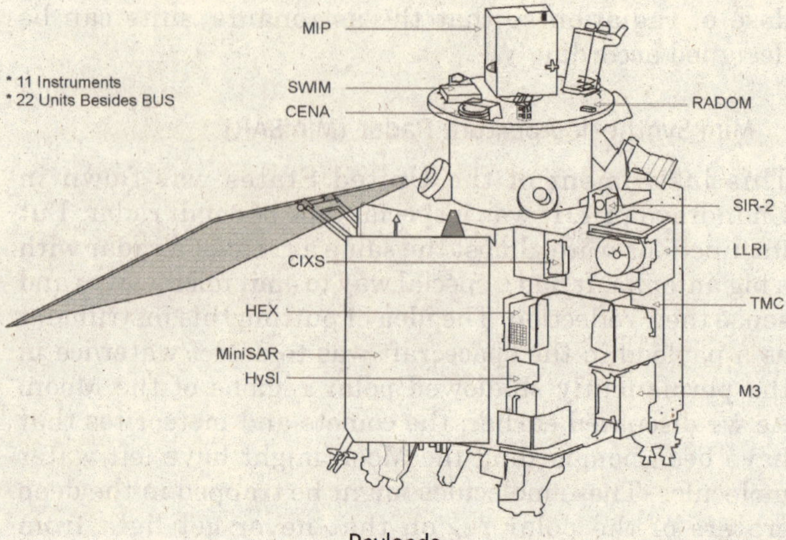

Payloads

with sensitive instruments. In addition, by agreeing to fly instruments from other countries, this mission set an example of how international cooperation can be an important part of space exploration.

THE ROCKET

Surprising though it may seem, Indian rocketry was born in a church. This church is like any other that you may see in Kerala. But it has a long history. It was founded by Saint Francis Xavier in 1544. To start with, the church was a shed thatched with dried coconut leaves. The fisherfolk from the coastal village of Thumba used to gather in the shed to listen to the Bible teachers. Once a year, a priest from Thiruvananthapuram came to the church to administer sacraments.

THE MAIDEN EFFORT

ISRO's space programme started with the launching of sounding rockets. The idea was to take up scientific investigation of the upper atmosphere above the geomagnetic equator. Since this passes through Thumba village, it was necessary to have the programme at Thumba. The only suitable building available in the village was the old church, and that was where the programme started. Today the church building houses a space museum.

The first sounding rocket took off in November 1963. The zeal with which ISRO's engineers built and flew these rockets was missionary. It was like the zeal of the Bible teachers in the sixteenth century who braved the weather to preach in the thatched church. Subsequently, ISRO's engineers built and launched more advanced rockets with the same enthusiasm.

THE FATHER OF ROCKETRY

You would have seen the small rockets that are fired on festival days in firework displays. Large rockets that launch satellites/spacecraft are based on the same physical principle as these small rockets. The principles of the modern-day rocket were first developed by Konstantin Tsiolkovsky of Russia.

Tsiolkovsky was deaf from his childhood. But he worked hard and studied mathematics and physics. By applying laws of physics, he improved the design of rockets. He was the one who suggested the idea of staged rockets, that is, a number of rockets kept on top of one another. Tsiolkovsky showed that travelling in space was possible using staged rockets. Because of his work on rockets, he was known in Russia as the 'father of space travel'.

WE OWE IT TO GODDARD TOO

It was Robert Goddard of America who built the world's first liquid-fuelled rocket in 1926. The rocket used petrol and liquid oxygen as fuel. In 1929, he used the rocket for the first time to measure pressure and temperature at different heights in the atmosphere. He kept a barometer, a thermometer and a camera in the nose cone of the rocket. When the rocket climbed up to the upper atmosphere, the nose cone opened and the readings of the barometer and thermometer were recorded on the camera. When the rocket came down, the payload was recovered and the readings of pressure and temperature were obtained from the camera.

Short by Only 238,788 and Half Miles

Goddard was convinced that a rocket could go to the Moon. One of the rockets he tried to send to the Moon could go only half a mile up. People made fun of him saying that he missed the Moon by only 238,799 and half miles because the distance to the Moon is 238,800 miles. Goddard was known as the 'moon-rocket man' in USA.

HOW DOES A ROCKET WORK?

Newton's third law of motion explains how a rocket works. As a reaction to hot gas rushing out from the nozzle with high speed, the rocket goes up in the opposite direction.

As the rocket climbs higher, fuel is consumed and ejected out as gas from the nozzle. As a result, the mass of the rocket decreases continuously. In terms of Newton's second law of motion, the rocket is accelerated continuously. The fuel in the rocket burns at a constant rate and this produces force at a constant rate. A rocket's performance

Newton's Laws of Motion

- A body persists in its state of rest or of uniform motion unless acted upon by an external unbalanced force.
- Force equals mass times acceleration: the net force on an object is equal to the mass of the object multiplied by its acceleration.
- To every action there is an equal and opposite reaction.

is decided by the thrust produced by its fuel. The parameter for this is the specific impulse. Specific impulse is defined as the thrust produced per unit weight of fuel consumed per second.

The mass ratio

An important factor that determines the efficiency of a rocket is its mass ratio. This is the ratio between the total mass of the rocket before it takes off and the mass after the fuel is consumed. Higher this mass ratio, higher would be the efficiency of the rocket.

BUILDING A ROCKET

One kind of energy is converted to another kind in a rocket. In a chemical rocket, the chemical energy of the fuel is converted into kinetic energy. Most of the rockets that we have today use either solid fuel or liquid fuel or both. In a solid fuel rocket, the fuel and oxidizer are mixed into a solid form. In a liquid fuel rocket, the fuel and the oxidizer are in liquid form and kept separately. The combination of the fuel and oxidizer is called 'propellant'.

Parts of a rocket

The combustion chamber, nozzle and nose cone are important parts of a rocket. The propellant burns in the combustion chamber producing hot gas at high pressure. The nozzle expands the gas and produces thrust. The top portion of the rocket is called the nose cone. The payload is kept in the nose cone. When the rocket goes up, the outer surface of the nose cone gets heated up because of friction with the atmosphere. To protect the payload, the outer surface of the nose cone is made of special materials and is called the heat shield.

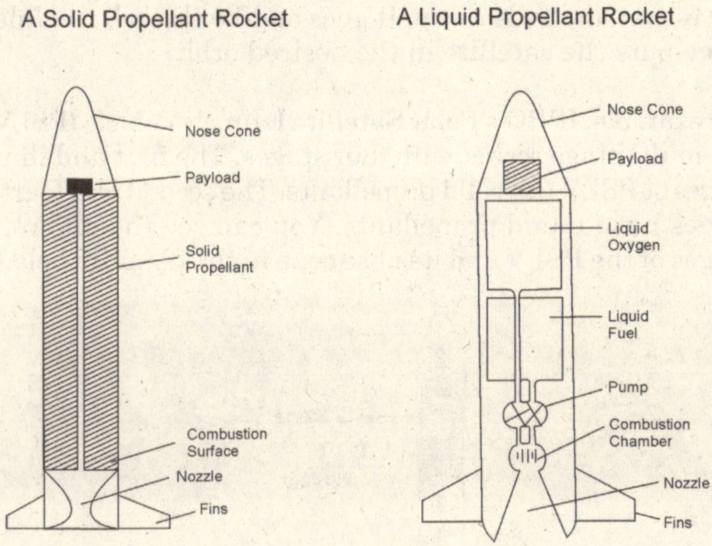

Parts of a Rocket

The staged rocket

A rocket with a single stage is not used for launching a satellite because it would not have the required velocity. For this a multi-stage rocket is used. The lowest part of the multi-stage rocket is called the first stage. It has the maximum propellant and can produce the greatest thrust. Strap-ons are attached to the first stage. Strap-ons are the additional rocket boosters that provide extra energy to the vehicle and thereby increase the overall capability of the vehicle. The successive stages produce lower thrust.

When a multi-stage rocket goes up, the propellant of the first stage takes the rocket to a certain height. When the propellant is used up, the stage separates from the rocket and falls away. At the same time, the propellant of the second stage is ignited and it takes the rocket further up. After a specified time the second stage also separates from the rocket and falls away. It goes on like this till the final stage puts the satellite in the desired orbit.

For example, ISRO's Polar Satellite Launch Vehicle (PSLV) is a multi-stage rocket with four stages. The first and third stages of PSLV use solid propellants. The second and fourth stages have liquid propellants. You can see the different stages of the PSLV and its nose cone in the diagram below.

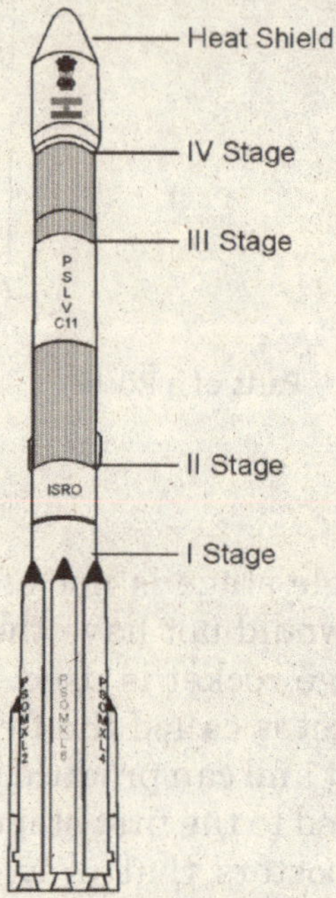

PSLV

ISRO'S MAGIC BOX OF ROCKETS

Sounding Rockets

ISRO's rocketry programme started with sounding rockets. On 21 November 1963 a sounding rocket took off from Thumba village, putting the obscure fishing village on the world map. ISRO has flown many sounding rockets since then either from Thumba or Sriharikota in Andhra Pradesh.

These sounding rockets are used to carry out scientific experiments in the upper atmosphere. ISRO's sounding rockets go up to a height of 50 to 550 km, carrying payloads that weigh 10 to 100 kg.

Satellite Launch Vehicle-3 (SLV-3)

SLV-3 was India's first launch vehicle to put a satellite in space on 18 July 1980.

A.P.J. Abdul Kalam was the man in charge of designing, developing and launching SLV-3. It was a small launch vehicle by world standards. It weighed 17 tonne at lift-off. It was 22 metre long and measured 1 metre in diameter. Though tiny, SLV-3 gave India the ticket to enter the big league of spacefaring nations.

Augmented Satellite Launch Vehicle (ASLV)

SLV-3 could carry a satellite of 40 kg. It was certainly not the vehicle to carry orbiting operational satellites to space. By 1980 ISRO was already thinking of designing and building a vehicle which could carry a satellite of 1,000 kg to a height of 900 km. But it was too big a jump from SLV-3. So ISRO designed and built a vehicle as a technological bridge between the two. That was the ASLV.

ASLV weighed 39 tonne at lift-off and could carry a satellite of 110 kg into space. ASLV flew successfully twice. ISRO's experience with it was very instructive. It was a low-cost platform on which ISRO could prove most of the crucial technologies needed for operational launch vehicles. In that sense, ASLV was a watershed in ISRO's history of launch vehicles. If ISRO learnt the alphabet of launch vehicle design through SLV-3, it learnt the grammar through ASLV.

Polar Satellite Launch Vehicle (PSLV)

The PSLV was bigger than SLV-3 and ASLV. It was designed to carry a satellite of about 1,000 kg to a height of 900 km. But, over a period of time, the capability of PSLV has increased. It can now carry a 1,600 kg satellite.

While SLV-3 got India a place in the community of spacefaring nations and ASLV gave India the rites of passage into launch technology, PSLV announced to the world that India had arrived. PSLV is a truly world-class launch vehicle. It had fifteen successful flights in a row. It has so far carried thirty-eight satellites to different orbits.

A World Record

PSLV can now carry several satellites in a single flight and launch them to different orbits. In January 2007, PSLV-C7 carried four satellites including two foreign satellites. In April 2008, PSLV-C9 created a world record by carrying ten satellites in a single flight.

Geosynchronous Satellite Launch Vehicle (GSLV)

PSLV is a versatile and reliable rocket. But, even with an operational PSLV, there was something missing. ISRO had no muscle to climb to 36,000 km carrying its communication satellites. GSLV did that in 2001, making India the sixth country to have that capability. GSLV can fly to a height of 36,000 km, carrying a satellite of about 2,000 kg.

Geosynchronous Satellite Launch Vehicle-Mark III (GSLV-Mk III)

Right now ISRO is developing a new vehicle called GSLV-MkIII. It can launch a satellite weighing 4,000 kg to the Geostationary Transfer Orbit.

THE ROCKET FOR THE MOON

There was a lot of discussion in ISRO about choosing an appropriate rocket for the Chandrayaan 1 spacecraft. Finally PSLV was selected. This was for two reasons. One is that PSLV is a reliable vehicle. By 2008, PSLV had successfully launched twenty-nine satellites to a variety of orbits. The other reason is that PSLV has a strong heritage from Kalpana 1. It was PSLV-C4 that had launched Kalpana 1, a weather satellite, into the Geosynchronous Transfer Orbit on 12 September 2002.

PUTTING ON WEIGHT

PSLV was originally designed to launch satellites of about 1,000 kg weight into an orbit of 900 km. The first successful flight of PSLV carried a satellite of 850 kg in October 1994. Since that time, the weight that PSLV can carry has gradually improved from 850 kg to 1,600 kg. The improvement has been achieved through imaginative means like loading more propellants in the stage motors and strap-ons, using stronger but lighter material for the satellite mounting structure and changing the sequence of firing the strap-on motors.

THE CHOSEN ONE

PSLV-C11 was chosen to carry Chandrayaan 1 spacecraft. It is a 44.4 metre tall vehicle with four stages. While the first and third stages have solid propellants, the second and fourth stages have liquid propellants.

The first stage carries 138 tonne of propellant. It is one of the largest solid propellant boosters in the world. There are six solid strap-on motors strapped on to the first stage. Each of these strap-on motors carries 12 tonne of solid propellant.

The second stage carries 41.5 tonne of liquid propellant.

The third stage uses 7.6 tonne of solid propellant. The fourth stage has a twin-engine configuration with 2.5 tonne of liquid propellant. At the top of the PSLV-C11 vehicle, there is a bulbous fairing. It has a diameter of 3.2 metre. It protects the satellite and is discarded after the launch vehicle clears the dense atmosphere.

THE LABYRINTH

A rocket engine is a very complicated thing. But there are other far more complicated systems in PSLV-C11 apart from its rocket engines. For example, there are systems that perform the important task of separating the stages from the vehicle after they complete their work. A complex telemetry system gives the information whether the rocket is functioning properly during its journey. PSLV-C11 has an electric brain with six microcomputers, which controls the flight without any help from outside.

THE MEN WHO BUILT IT

PSLV-C11 was designed and built by the ISRO engineers at Vikram Sarabhai Space Centre at Thiruvananthapuram. It is ISRO's lead centre for development of launch vehicles and sounding rockets. PSLV-C11 was launched from Satish Dhawan Space Centre, Sriharikota, Andhra Pradesh.

THE GROUND STATION

If you travel from Bangalore to Mysore by road, one of the turnings to the right will take you to the Big Banyan Tree. The tree is very old and has hundreds of roots supporting the leaf canopy like poles in a tent. The tree is also famous for the large number of monkeys it houses. The monkeys swing from the roots as if performing in a circus. If you make the mistake of going near the monkeys, they will rush at you, walking on their flat hands like legless beggars and making menacing noises and funny faces.

Beyond the Big Banyan Tree is the sleepy village of Byalalu with tiled houses sitting cheek by jowl as if carrying on a whispered conversation. Rising tall over the tiled roofs are two huge dish antennas, looking determinedly at the sky. These are the antennas built by ISRO for the ground segment of the Chandrayaan 1 mission.

WHAT DOES THE GROUND STATION DO?

The ground station keeps in touch with the orbiting spacecraft. It has large antennas, radio receivers, radio transmitters, tape recorders and computers. It is very important for the ground station to be in radio contact with the spacecraft in the orbit. While Chandrayaan 1 was orbiting the Moon, the radio signals it sent out provided telemetry data. The information that the radio signals provided was very detailed. The amount of information given every day was so large that together they could fill many large books.

The telemetry data had two parts. The first part gave what the payloads of the spacecraft were seeing from the orbit. This is called observational data. The second part provided what is called the house keeping data. This data gives information about the performance of the different sub-systems of the spacecraft such as temperature, availability of power and direction of the spacecraft. House keeping data is important because it tells us about the health of the satellite.

Ground stations are also the place from where radio commands are given to an orbiting spacecraft. These commands are in the nature of instructions to the spacecraft to perform various tasks. The tasks include functions such as switching certain equipments on or off, firing the thrusters, and firing the liquid apogee motor for controlling or changing the orbit of the spacecraft. The ground station is also used for tracking purposes. This is to determine the orbit of the spacecraft.

SPACE TRACKING AND DATA SYSTEMS

The great success of space programmes all over the world during the last six decades would not have been possible without effective space tracking and data systems. These support systems provided several key inputs: position of the spacecraft in space, scientific data from the spacecraft, engineering data to monitor the performance of the spacecraft, and transmission of commands from the ground to the spacecraft.

As space programmes have increased in scope and complexity, so have the tracking and data system networks.

The quantities of data transmitted by today's spacecraft are thousand times greater than those in the 1950s. For example, when Explorer 1, the first successful satellite of the US, was launched in the 1950s, its orbit was plotted on the basis of data gathered by portable radio tracking stations in Nigeria, Singapore and California. When the simple spacecraft missions of the early days gave way to complex missions, it became necessary for the US to establish the Deep Space Network that could cater to the requirements of deep space missions. Today, the US is the world leader in development of antennas and low-noise receivers, tracking, telemetry and command systems, digital signal processing and deep space navigation.

DEEP SPACE NETWORK

The Deep Space Network (DSN) of the US consists of three deep space communication facilities located around the world but placed 120 degrees apart. They are:

- The Gladstone Deep Space Communications Complex, California
- The Madrid Deep Space Communications Complex, Spain
- The Canberra Deep Space Communications Complex, Australia

Each of these facilities is situated in mountainous terrain which is shaped like a bowl. Such a location acts as a shield and protects the facility from radio frequency interference. The location also makes it possible to observe the spacecraft at all times even as the Earth rotates.

Each facility has at least four deep space networks equipped with ultra-sensitive receiving systems and large parabolic dish antennas. They are:

- One 34 metre (111 feet) diameter High Efficiency antenna

- One or more 34 metre Beam Waveguide antennas
- One 26 metre (85 feet) antenna
- One 70 metre (230 feet) antenna

The large antennas of the DSN are used only during emergencies. In normal times, the spacecraft can operate on the smaller antennas. During emergencies, the spacecraft are forced to use less of the high-gain antennas located on the spacecraft. High-gain antennas is a technical term which means a powerful antenna where there is very little loss in communication. The most famous example is that of Apollo 13, when the limited battery power and inability to use the spacecraft's high-gain antennas reduced signal levels to a very low level, and the spacecraft was forced to use the DSN's big antenna located at Australia. In fact, the use of this big antenna saved the lives of the astronauts on Apollo 13.

ANTENNAS

The primary function of the antennas is to relay communication to and from the spacecraft. They are a kind of conductor by which electromagnetic waves are transmitted or received. An ordinary antenna unfurls like a giant umbrella when deployed. The antenna's primary reflector surface is a gold-clad wire-mesh, woven like cloth on the same type of machine used to make material for hosiery. When deployed, the antenna's mesh is stretched tautly on supporting ribs by fine thread-like cords and looks like a glittering spider web made of metal.

INDIAN DEEP SPACE NETWORK (IDSN)

ISRO built a deep space network for the Chandrayaan 1 mission, consisting of a 18 metre dish antenna (IDSN-18) and a 32 metre wheel and track antenna (IDSN-32), both located in Byalalu.

The deep space network was necessary because the Chandrayaan 1 spacecraft was 4,00,000 km away, and the radio waves from the spacecraft become very weak by the time they travel the distance to Earth. They are so weak that it becomes difficult to retrieve the valuable information they carry. That is why big dish antennas with sensitive equipment were necessary to recover the information in those radio waves after separating them from the background electronic noise.

INDIAN SPACE SCIENCE DATA CENTRE (ISSDC)

ISRO also built the ISSDC, a new facility for handling the scientific information sent by the Chandrayaan 1 spacecraft. ISSDC stores the information for future use and distributes the information to scientists and general public.

THE NETWORK

There are several ISRO stations located at Bangalore, Lucknow, Mauritius, Bearslake, Biak, Brunei, Thiruvananthapuram and Port Blair which were used for looking at Chandrayaan 1 at its launch and during its orbit phase and normal operations. The support of external network stations at Gladstore, Canberra, Madrid, Hawaii, Alcantara and Cuibá was also taken.

The Spacecraft Control Centre, located at the ISTRAC campus of ISRO at Bangalore, was the nerve centre for Chandrayaan 1 spacecraft operations. It was responsible for all the operations of the spacecraft during various phases of the mission.

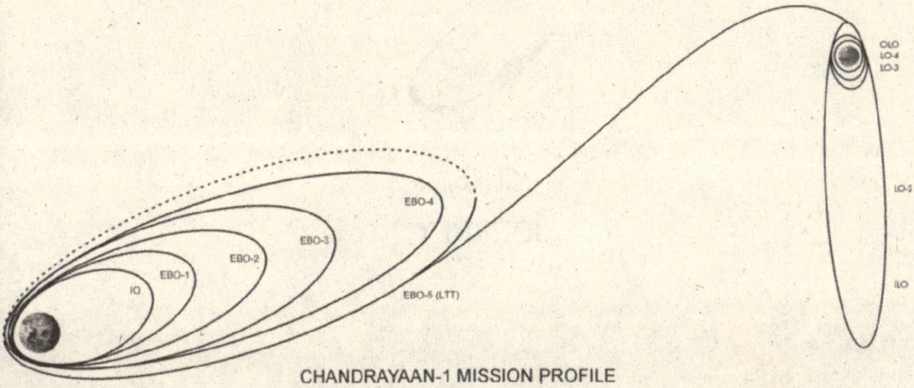

CHANDRAYAAN-1 MISSION PROFILE

The Flight Path

The diagram above shows the path of the Chandrayaan 1 spacecraft to the Moon. The ground station guided the spacecraft along this path. The ground segment of the mission performed the important task of receiving radio signals sent by the spacecraft. It also transmitted radio commands to the spacecraft during various phases of the mission.

THE JOURNEY

Mornings come early to Sriharikota. It is an island in the Nellore district of Andhra Pradesh. The island is shaped like a spindle. On one side is the Bay of Bengal and on all other sides, it is encircled by Pulicat lake. When the rains fill up the lake during the north-east monsoon, thousands of migratory birds visit the island from faraway places, turning Sriharikota into a paradise for bird watchers. It is from Satish Dhawan Space Centre in Sriharikota that ISRO's satellites and spacecraft are launched.

The Launch Pad

Turning off from Sullurupeta, a small town on the national highway that connects Chennai and Kolkata, a twenty-minute drive towards the east on the road laid across Pulicat lake takes one to Sriharikota. The first flight of Rohini 125, a small sounding rocket, had taken place from Sriharikota on 9 October 1971. Since then, the facilities have been expanded to meet the growing needs of ISRO.

On 22 October 2008, hours before the launch, PSLV-C11 stood proudly at Second Launch Pad, fully decked up like a bride and waiting to be launched. It was a sight that inspired awe. The giant rocket, at 44 metre, was as tall as a high-rise building fifteen storey high, and weighed 320 tonne.

The Second Launch Pad is an engineering marvel. What we call a launch pad is known as 'Umbilical Tower' in rocket engineer's jargon. It was in front of the 83 metre tall Umbilical Tower that PSLV-C11 stood, on its Mobile Launch Pedestal. Down below was a huge pit that was about five storey deep. This pit—the flame pit—was to divert the hot exhaust gases that keep escaping from the nozzle of the rocket. Without this pit, the hot exhaust gases would bounce back into the rocket and do severe damage to the vehicle.

The Umbilical Tower has several huge arms that move vertically as well as horizontally. This is to enable the ground personnel to service it. Many electrical cables, propellant and gas supply lines, and the pipe supplying cool air to the spacecraft to keep it comfortable, hung like giant umbilical cords. Four lightning towers which were even taller than the Umbilical Tower stood around the vehicle as if in a protective shield. It was a majestic sight to behold.

THE COUNTDOWN

. . . Countdown . . . . . .

A launch vehicle has several rocket stages which come to life one after the other. There are the strap-ons—looking like candlesticks—which are ignited along with the first stage of the launch vehicle. There are also a number of critical activities like the filling of liquid propellants, pressurization of the propellants, arming of various circuits and calibrating of the gyroscopes, which have to be done in an assigned time prior to the launch.

These precisely timed and sequenced events which take place before the launch vehicle lifts off are referred to as 'final countdown'. During these activities time is counted back to zero, when the main stage is ignited. The countdown begins anywhere between 100 to fifty hours before the main engine is ignited. The countdown is important because it provides time to solve any problem that might crop up unexpectedly at the last minute.

The unfamiliar strap-ons

The countdown started fifty-two hours before the launch. PSLV-C11 was comfortably ensconced in the launch pad, but ISRO scientists were keeping their fingers crossed. They were nervous about several things. One was the use of larger strap-ons in PSLV-C11. This was being done for the first time. The PSLV-C11 had six of them to provide additional punch to the vehicle at the time of its lift-off. Would the strap-ons behave as expected?

The tight launch window

Another cause for anxiety was the nature of the mission. The launch window (the most suitable interval of time for the launch for a safe and efficient journey) was rather tight. This was a real challenge because both the place of launch (Earth) as well as the ultimate target (Moon) were speeding away in space and Chandrayaan 1 had to keep its appointment with a moving target at a distance of nearly 4,00,000 km at a precise time and place.

The treacherous weather

The third worry was the weather. The north-east monsoon is a regular visitor to Sriharikota which is on a coastal strip of land. It often rains there during October and on to December. ISRO scientists knew this but they had taken a calculated risk in fixing the launch date of Chandrayaan 1 on 22 October 2008. As ill luck would have it, there was heavy, intermittent rain in Sriharikota during the final few days leading to the launch.

Why was October 22 Chosen as the Launch Date?

An earth orbiting satellite is launched during a particular time interval called 'launch window'. This launch window is the interval of time during which the satellite can be launched efficiently from the point of view of propulsive capabilities of both the launch vehicle as well as the satellite's own rocket engine. In the case of a spacecraft like Chandrayaan 1 that was venturing beyond the earth orbit, the launch window is much tighter. This is because the launch pad (the Earth), the target (in case of Chandrayaan 1, the Moon) and the spacecraft, all will be moving in space at tremendous speeds. Chandrayaan 1 was scheduled to be ready for launch in October 2008. After the spacecraft's launch, about two weeks would be needed to take it near the Moon. Calculations of orbital mechanics showed that there would be two opportunities for Chandrayaan 1 in November 2008 during which a 'Lunar Orbit Insertion' manoeuvre could be performed. The first one was on 8 November 2008. It was, therefore, decided to launch Chandrayaan 1 on 22 October 2008 so that the spacecraft would have enough time to reach the vicinity of the Moon for the lunar orbit insertion manoeuvre to be performed on 8 November 2008.

In the early morning hours of 21 October 2008, thunder rumbled ominously. Even by afternoon, there was no improvement in the weather. As midnight approached,

Sriharikota had heavy rains. With no sign of the rain easing, the ISRO scientists went into a huddle.

Unmindful of the pouring rain, an enormous crowd had started gathering in Sriharikota to witness the launch. There was a steady flow of media persons from Chennai throughout the night. Journalists from Delhi and abroad also started arriving. The historic occasion had drawn thousands of people to Sriharikota, turning the area into the largest-ever camp site.

Countdown Timer

The ticking of the digital countdown clock causes a great deal of excitement. Through the final few minutes before the main engine is ignited (referred to as 'T Zero' by rocket engineers), when the detailed health check-up of the rocket is performed by computers, the adrenaline level in the Mission Control Centre as well as the Launch Control Centre shoots up and an eerie silence prevails. What one hears at the time are the cryptic announcements by a person known as the 'countdown timer'. These announcements declare: 'stand by the time mark', 'mark five minutes and counting', 'mark three minutes and counting', 'that's thirty seconds'. When the countdown timer announces, 'ten, nine, eight, seven, six, five, four, three, two, one', the excitement reaches a crescendo. A moment later, as soon as the word 'zero' is announced, the main engine comes to life and builds sufficient thrust for lift-off. The countdown ends, only to be followed by a more exciting countup during which critical flight events like stage separation and ignition of the successive stages take place. Countup culminates in the successful injection of the spacecraft into the required orbit. With this, the spacecraft begins its life in space.

From tension to jubilation

As the countdown clock ticked away, the tension was palpable in the Mission Control Centre, the nerve centre of all the operations that went on in the island. The Mission Control Centre, with its rows of consoles, was a flurry of activity. Particularly interesting was the display and control system, a series of large projection screens on the front wall for a variety of displays ranging from charts that plot the vehicle's journey to display of dynamic or constantly changing data showing the altitude and velocity of the moving vehicle.

ISRO chairman Madhavan Nair and other senior ISRO scientists were at their consoles in the Mission Control Centre busy monitoring the launch operations. As if by a miracle, the weather gods had relented and except for an intermittent drizzle, there was no rain. But the scientists were still worried about the thick clouds, powerful winds and lightning. At the Launch Control Centre down below, engineers who had built the PSLV-C11 were glued to the computer screens that gave information about the health of the rocket. As the countdown clock ticked away, the pre-launch operations proceeded without a hitch.

Twelve minutes before the launch, the computer took over the task of checking hundreds of health parameters of PSLV-C11 that stood on the Second Launch Pad about 6 km away. As the voice of the countdown timer announced 'T minus one minute', the excitement, expectation, apprehension and tension peaked. The countdown reached 'zero' and the giant first stage of the vehicle together with four of the six strap-ons ignited almost simultaneously in a bright flash of blinding light. PSLV-C11 lifted off at 6.22 a.m., and it was on schedule. It took off majestically, soaring above the tall eucalyptus trees. It was a blinding pillar of flame silhouetted against the early morning sky. The deafening roar of the powerful engines boomed like a million guns all firing at the same time. In a matter of seconds, the vehicle disappeared in a blaze of glory into the clouds, leaving behind trails of orange fumes that rose in lazy spirals. The vehicle was on target. After 109 seconds, the first stage separated. The vehicle had reached an altitude of 62 km. The atmosphere in the Mission Control Centre changed from one of expectation to that of jubilation. 'The tension which prevailed till then, suddenly vanished, only to be replaced by another similar one as the vehicle rose into the cloudy sky,' recalled S.K. Shivakumar, who heads ISTRAC.

The second stage takes off

The second stage of PSLV-C11 had been ignited and the vehicle was climbing. At a height of about 116 km, the payload fairing separated. It meant that the vehicle had cleared the dense atmosphere. The second stage separated at 245 km. It had taken 266 seconds from the time the vehicle had lifted off.

The third stage takes over

The third stage had been ignited now and the vehicle was climbing steadily. At a height of 577 km, and 488 seconds after lift-off, the third stage separated.

The fourth stage is ignited

Now the fourth stage had been ignited. This is the stage with the twin-engine configuration with 2.5 tonne of liquid propellant. After the vehicle reached a height of 820 km, the fourth stage was cut off. It had taken exactly 993 seconds from the lift-off. The rocket had burnt itself out but it had done its job. The time it had taken, the height it had reached and the velocity it had achieved were as planned. It was a perfect launch. There was thunderous applause all around the Mission Control Centre.

GOING AROUND EARTH

The spacecraft had now reached an orbit that resembled a highly elongated circle round the Earth. As the spacecraft went around Earth in this initial orbit, the point that was of minimum distance from Earth (perigee) was at a height of 255 km. The point of maximum distance (apogee) was at 22,860 km. In this orbit, Chandrayaan 1 orbited Earth once in about six and half hours.

23 October 2008

The Liquid Apogee Motor (LAM) of the spacecraft was fired at 9 a.m. when the spacecraft was near the perigee. The firing was done to raise the apogee to 37,900 km and the perigee to 305 km. The spacecraft now took eleven hours to go around Earth once.

26 October 2008

LAM was fired again at 7.48 a.m. for about nine and a half minutes. With this, Chandrayaan 1 entered a much higher elliptical orbit around Earth. The apogee of this orbit was at 1,64,600 km and the perigee at 348 km. In this orbit, Chandrayaan 1 took about seventy-three hours to go around Earth once. The spacecraft had now entered deep space after crossing 1,50,000 km distance from the Earth.

29 October 2008

The orbit of the spacecraft was raised again at 7:38 a.m. by firing LAM for about three minutes. With this, Chandrayaan 1 entered into a more elliptical orbit whose apogee was at 2,67,000 km and the perigee at 465 km.

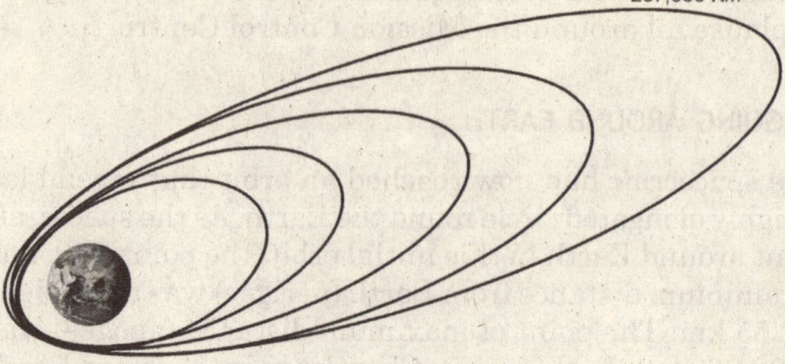

Halfway to the Moon

The spacecraft had now travelled more than halfway to the Moon.

The health of the spacecraft was being continuously monitored from the Spacecraft Control Centre at ISTRAC in Bangalore with support from the Indian Deep Space Network antennas at Byalalu. All systems onboard the spacecraft were performing normally.

31 October 2008

The Terrain Mapping Camera (TMC) onboard the spacecraft was switched on through a series of commands issued from the Spacecraft Control Centre. An analysis of the first image received by the Indian Deep Space Network at Byalalu and later processed by the Indian Space Science Data Centre showed the excellent performance of the camera. The first image taken at 8 a.m. from a height of 9,000 km showed the northern coast of Australia.

4 November 2008

The fifth and final orbit raising of Chandrayaan 1 was done at 4.56 a.m. by firing LAM for about two and a half minutes.

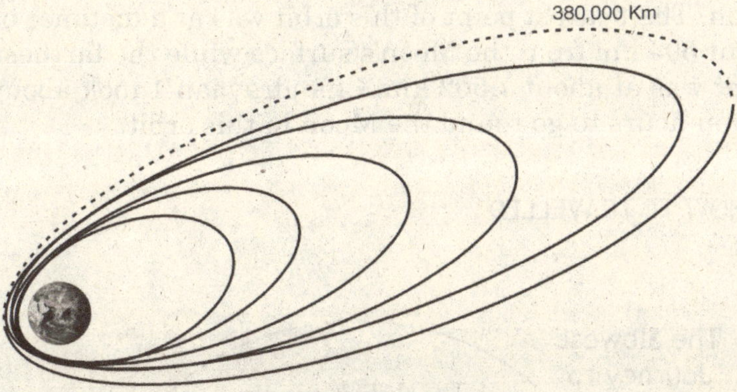

Lunar Transfer Trajectory

With this, Chandrayaan 1 had entered the Lunar Transfer Trajectory with an apogee of about 3,80,000 km.

ENTERING THE MOON'S ORBIT

8 November 2008

On this date Chandrayaan 1 entered the Moon's orbit. This was the first time that an Indian spacecraft had broken away from Earth's gravitational field and reached the Moon. This historic event occurred after the spacecraft's LAM was fired at 4.51 p.m. for 817 seconds. The complex Lunar Orbit Insertion Manoeuvre was performed by the Spacecraft Control Centre. The Indian Deep Space Network supported the crucial task of transmitting commands and monitoring the event continuously.

The LAM was fired when the spacecraft passed at a distance of about 500 km from the Moon to slow it down. You may ask: why slow it down? It was to allow the gravity of the Moon to capture the spacecraft into an orbit around the Moon. The spacecraft was now orbiting the Moon in an elliptical orbit that passed over the polar regions of the

Moon. The nearest point of this orbit was at a distance of about 504 km from the Moon's surface while the farthest point was at about 7,502 km. Chandrayaan 1 took about eleven hours to go round the Moon in this orbit.

HOW IT TRAVELLED

The Slowest Journey to the Moon

The slowest mission to reach the Moon was SMART 1, a probe launched by the European Space Union. It was launched on 27 September 2003. It arrived at its destination on 11 November 2004. It had taken one year, one month and two weeks. But the mission had been very fuel efficient. It had used only 82 kg of propellants for the entire journey. That is because it had used an ion engine to take it to the Moon.

ALMOST THERE

10 November 2008

Chandrayaan 1 reached its intended operational orbit at a height of 100 km from the surface of the Moon. This happened after a series of three orbit-reduction exercises carried out by repeatedly firing the LAM. As a result of these exercises, the farthest point of Chandrayaan 1's orbit from the Moon's surface was reduced from 7,502 km to 255 km, and finally to 100 km. The nearest point was reduced from 200 km to 182 km and finally to 100 km. In this orbit Chandrayaan 1 took about two hours to go round the Moon once.

From this operational circular orbit of 100 km height passing over the polar regions of the Moon, the spacecraft would do chemical, mineral and photo-geological mapping of the Moon with its eleven scientific payloads. Two of these payloads, Terrain Mapping Camera (TMC) and Radiation Dose Monitor (RADOM) had already been switched on and were working well.

HOISTING THE TRICOLOUR

14 November 2008

The Indian national flag was placed on the Moon's surface on Pandit Jawaharlal Nehru's birthday. Modern Indian space programme was initiated in 1962 when Pandit Nehru was the Prime Minister of India. The flag was painted on the sides of the Moon Impact Probe (MIP). The MIP hit the lunar surface at 8.31 p.m. (Indian Standard Time). It was the first Indian built object to reach the surface of the Moon. The point where the MIP hit the Moon's surface was near the south polar region.

Weighing 34 kg, the box-shaped MIP carried three equipment: a video imaging system, a radar altimeter and a mass spectrometer. The video imaging system took pictures of the Moon's surface as the MIP approached it. The radio altimeter measured the rate of descent of the MIP to the lunar surface. The mass spectrometer studied the extremely thin lunar atmosphere.

The MIP's journey to the Moon's surface began when it separated from Chandrayaan 1 at 8.06 p.m. (Indian Standard Time). The journey lasted twenty-five minutes. Information from its instruments was radioed to Chandrayaan 1 by the

MIP. The spacecraft recorded this in its onboard memory for later readout. Finally, the MIP had a hard landing on the Moon that terminated its functioning.

16 November 2008

Lunar Laser Ranging Instrument (LLRI), one of the scientific payloads, was turned on. This was when the spacecraft was passing over the western part of the Moon's visible hemisphere. Preliminary assessment of data from LLRI showed that the performance of the instrument was normal. LLRI would have to be kept continuously on to take 10 measurements per second of the day and night sides of the Moon.

The remaining seven scientific payloads were commissioned sequentially, and by the end of the year 2008 all of them had been switched on successfully. The exploration of the Moon with the full array of scientific instruments onboard Chandrayaan 1 had begun.

India now became the fifth country in the world to have sent a spacecraft to the Moon. The other countries that have sent spacecraft to the Moon are the United States, former Soviet Union, Japan and China. The European Space Agency, a consortium of seventeen countries, has also sent a spacecraft to the Moon.

THE FIRST RESULTS

Earth's satellite, the Moon, now had a satellite of its own—Chandrayaan 1! With the payloads beginning their work, a wealth of information started trickling in.

MEASURING RADIATION

On the very first day of the spacecraft in space, RADOM had been switched on. RADOM has provided very valuable information so far. The information is about the radiation environment of near Earth space and also that of Van Allen radiation belts surrounding Earth. As the spacecraft travelled deeper into space, RADOM acquired information on the radiation environment of deep space. Having been powered on continuously, RADOM provided a wealth of information about the radiation environment near the Moon till the mission was active.

Van Allen Radiation Belts

On 31 January 1958, the US Army sent a small satellite, weighing 15 kg, called Juno 1 which provided the American physicist James Van Allen with the first great discovery of the space age: the innermost of several radiation belts encircling Earth. These radiation belts are known as Van Allen radiation belts.

THE CAMERA IS CLICKING

The Terrain Mapping Camera (TMC) was switched on when Chandrayaan 1 was on its way to the Moon. This camera took 3D black and white pictures in which features as small as 18 feet are visible. While the spacecraft was still in Earth's orbit, the camera took pictures of Earth as well as the faraway Moon.

Once the Chandrayaan 1 spacecraft settled down to orbiting the Moon, the camera sent stunning pictures of the Moon. Thousands of such pictures have been received. When viewed through the anaglyph filter (special kind of glasses used to view 3D pictures), the height and depth of the Moon's terrain are so strikingly visible that we get the feeling that we are in a cavern full of craters.

THE 64-BAND IMAGER

HySI was the other Indian camera onboard the spacecraft. It took colour pictures and that too in sixty-four bands. The images taken by this camera over different periods of times are being studied by scientists.

THE X-RAY MACHINE

C1XS was developed jointly by UK and India. It used X-ray technique for measuring the availability of magnesium, aluminium, silicon, iron and titanium on the surface of the Moon. Though this instrument was switched on late, it detected the presence of magnesium, aluminium and silicon on the lunar surface. The instrument proved to be more sensitive than its designers thought it would be.

MOON MINERALOGY MAPPER (M3)

The American instrument was designed to assess and map mineral resources on the Moon. It detected the presence of the iron-bearing mineral pyroxene in the Orientale basin on the Moon.

More importantly, M3 has detected water in molecular form on lunar surface. M3 followed a process in which light is reflected in different wavelengths off different minerals. Scientists use these differences to find out what is present in the thin layer of the upper soil. This technique is known as reflective spectrometry. M3 has detected wavelengths of light reflected from the lunar surface that indicates a chemical band between hydrogen and oxygen. Since water is two hydrogen atoms banded to one oxygen atom (H_2O), this indicates the presence of water on the Moon.

MINI SYNTHETIC APERTURE RADAR

This instrument was looking for water ice in the permanently shadowed craters of the Moon's poles. These are parts of the Moon never seen before. Scientists in the US used the findings of the instrument to map and search the insides of the craters for water ice. Collection of data by the Mini Synthetic Aperture Radar and analysis of the data will help scientists to decide whether buried ice deposits exist in the permanently shadowed craters near the Moon's poles.

SIR-2 AND SARA

The remaining two foreign payloads (SIR-2 and SARA) were switched on. SIR-2 was the radiation band through which various minerals and ice reveal their existence. SARA studied atoms. Like the Indian payload HEX, these two instruments were switched on quite late.

THE LEGACY OF CHANDRAYAAN 1

For most people in India, the Chandrayaan 1 mission raised a million hopes. The amount of curiosity the mission aroused since its successful launch shows that even the common man is proud of the mission. Almost everything, from the manoeuvres it did to the planting of the tricolour on the Moon's surface and the pictures it sent back, electrified the entire nation.

Though Chandrayaan 1 mission came to an end earlier than expected with the loss of radio contact with the spacecraft on 29 August 2009, the mission was important in several ways. First, ISRO's success in designing, developing, building, testing and launching a lunar craft, putting it into an orbit around the Moon and sending a probe to lunar surface to impact in a designated area, was a spectacular achievement.

Second, ISRO was successful in developing an indigenous deep space network capable of communicating with and giving commands to a spacecraft orbiting at such a great distance. Third, the mission achieved precision positioning in the first attempt. Most first missions of the US, Soviet Russia and Europe either overshot the Moon or crashed into its surface.

And most important, Chandrayaan 1 achieved a historic first by discovering water on the lunar surface. Water was detected by the Moon Minerology Mapper, the American payload on board the Chandrayaan 1 spacecraft. The water detected on the lunar surface was in molecular form and in trace amounts, and that too on the uppermost layer. The thin sheet of water was found to be trapped a few millimetres under the lunar soil.

Chandrayaan 1 made this most significant discovery of all space expeditions just before its premature end. Water had not been found in the rocks and dust brought back by the Apollo and Luna missions. The Lunar Prospector had found evidence of hydrogen-bearing material in the north and south polar regions of the Moon. It was, therefore, assumed that except for the possible presence of water at the poles, there was no water on the Moon. Chandrayaan 1 has proved this assumption wrong by discovering water trapped a few millimetres under the soil.

Though only a molecular presence of water has been detected, Chandrayaan 1 had access to only the top few millimetres of lunar soil, and that too on the bright side of the Moon. The water detected might be the equivalent of one litre, but it is enough to fuel hopes of finding more water as ice in the darker, unexplored parts of the Moon.

This discovery of water opens up the possibility of setting up base camps on the Moon to carry out more detailed studies of its structure and processes. This means more exciting times ahead in lunar research and space studies. It also opens up the exciting possibility of using the Moon as a base for future human exploration of deep space.

The spacecraft mission may have come to a premature end, but the scientific mission of Chandrayaan 1 will continue. There has been a deluge of data from the spacecraft in the form of 3D colour pictures and radar images. The spacecraft has clicked about 70,000 pictures and has identified a large number of minerals. Radar imaging of the poles will tell us more about the presence of water. A detailed analysis of the data processed and stored at the Indian Space Science Data Centre in Byalalu will yield rich dividends and add to our understanding of the many aspects of the Moon.

The Indian payloads as well as those from abroad have worked to the satisfaction of their designers. Of the Indian payloads, TMC and the Hyperspectral imager have provided more than 70,000 pictures of the lunar surface. The Lunar Laser Ranging Instrument has helped scientists measure the height of surface features in lunar polar areas. The High Energy X-ray spectrometer has sniffed the X-rays emitted by heavy radioactive elements on the Moon.

Chandrayaan 1 imaging X-ray spectrometer (C1XS) performed better than expected and detected magnesium, aluminium and silicon on the Moon. The German SIR-2 has provided highly detailed information about the minerals on the Moon. The joint Swedish-Indian payload SARA covered the Moon many times and helped scientists understand the relationship between the Sun and a planet-sized body like the Moon without any magnetic field. The Bulgarian payload Radiation Dose Monitor was on from the day of launch and worked throughout the duration of the mission.

The two American payloads have also reaped rich data about the Moon. The Mini SAR payload studied the polar areas of the Moon that are in permanent darkness; and the Moon Mineralogy Mapper, in addition to discovering water, covered 95 per cent of the lunar surface and helped scientists understand the way Moon evolved.

Carle Pieters, one of the principal investigators of the Moon Mineralogy Mapper, says, the discovery of water 'opens a whole new avenue of lunar research and we have understood the physics of it to utilize it'. A statement by Brown University said, 'the discovery of water promises to reinvigorate studies of the Moon and potentially upend thinking of how it originated.'

Chandrayaan 1 was India's stepping stone into planetary exploration, and it faced immense challenges in the harsh lunar environment. Though its life was shorter than expected, it is a giant leap for India's space programme and has led to the biggest scientific discovery of the twenty-first century. India's maiden Moon mission has done us all proud.

PART THREE

THE FUTURE

WHAT NEXT?

The success of the Chandrayaan 1 mission has encouraged ISRO to take up more daring missions in the future. Chandrayaan 1 was an orbiting mission. Landing missions have some advantages that an orbiting mission does not have. For example, a landing mission can find definite answers to important questions about the Moon's core, availability of ice/water and the composition of the Moon. A landing mission can bring back samples from the high latitudes and far side of the Moon that can be studied for their chemical and mineral composition.

AN INDIAN ROBOT ON THE MOON

ISRO proposes to follow up Chandrayaan 1 with Chandrayaan 2 which will have a lander and two rovers. The idea is to develop robotic rovers that will take off from the orbiting spacecraft, land on the Moon, and rove over the Moon's surface. The rovers will analyse the Moon's surface and space environment. They will also look for helium-3. The data they collect will be transmitted to the orbiting spacecraft, which in turn will transmit to the ground station. For Chandrayaan 2, India has already signed an agreement with Russia to explore the Moon together.

Chandrayaan 2 will take off in 2011–12. It will be launched on the Geosynchronous Satellite Launch Vehicle (GSLV).

ISRO will make the orbiting satellite. There will be two robotic rovers, one made by ISRO and the other one by Russia. Scientific instruments of other countries will be placed in Chandrayaan 2.

A MISSION OF PEACE TO THE GOD OF WARS

Mars has always been of great interest to scientists. Just like Earth, Mars has ice caps on its poles, and some bluish ice clouds are visible away from the polar caps. There are also large dust storms on the surface of Mars. This is because storms of red dust whipped up by strong winds rage across its surface. The red dust has a lot of iron oxide which is nothing but rust.

The climate on Mars is hostile. The temperature does not rise above the freezing point even in summer and the land remains frozen for rest of the time. Would there be life in such a harsh place?

Life on Mars?

For almost 100 years now, scientists have been talking about the possibility of life on Mars. Pictures of Mars' surface show ridges on a canyon which would mean water flowed there in the past. These features could have been formed by free flowing water. There is strong evidence that Mars once had lakes and even floods of water. If there was water on Mars, could there have been life?

Gateway to the universe

Mars is also a stepping stone to the vast planetary world beyond. That is why ISRO's mission to Mars would be timely. The scientific objectives of ISRO's mission would be to understand the atmospheric processes on Mars, the weather and dust storms, the ionosphere, effect of solar

Water Flowed on Mars 1.25 Million Years Ago

Planetary geologists at Brown University, United States, have found a gully fan system on Mars that formed about 1.25 million years ago. It shows signs of water activity on the Red Planet. The fan offers evidence that it was formed by liquid water that originated in nearby snow and ice deposits. In a research paper that appeared in the 2009 March issue of *Geology*, the geologists of Brown University were able to date the gully system and give reasons for the presence of water.

wind, surface magnetic fields, and to search for water and surface resources.

A technology challenge?

Going to Mars is a challenge. ISRO is working on technology solutions for putting a spacecraft in low altitude orbit around Mars, and for developing instruments to monitor radiation, electric and magnetic fields, and energy particles. The mission to Mars would also require advanced command, communication, navigation and control.

A Treat?

Going to the Mars will be a real treat. We will get to see the biggest volcano in the solar system. It is called Olympus Mons. It is about 25 km high and 592 km wide. It is three times as tall as Mount Everest. Mars is named after the Roman god of war, but when ISRO goes there, it will go on a mission of peace!

MEETING THE SHOOTING STARS

There are a large number of small solar system bodies that revolve around the Sun in the area that extends between the orbits of Mars and Jupiter. These are called asteroids. The asteroids vary in size. They can be as small as a dust particle or as big as a small moon. Sometimes some small asteroids get knocked out of their positions and come towards Earth. These are called meteorites. These are metallic or stone bodies that fall on Earth or the Moon from outer space. You might have even seen them hurtling across the night sky. We call them shooting stars.

Asteroids and Earth

- On 30 June 1908 an asteroid exploded above Tungsuka, Siberia, with the force of a thousand atomic bombs. It created a colossal explosion in the air that flattened 2,000 square kilometre of forest land and destroyed 80 million trees.
- On 1 March 2009 an asteroid as tall as a ten-storey building and about 115 feet wide raced by Earth. The gap between the asteroid (now named as 2009 DD45) and Earth was just 72,000 km. This is one-fifth the distance between the Moon and Earth, and only twice the height of satellites in the geosynchronous orbit.
- The closest fly-by asteroid was 2004 FU 162—a small asteroid about 20 feet across—which came within about 6,500 km of Earth in March 2004.

On a collision course?

The area where these asteroids revolve around the Sun is called the Asteroid Belt. The Belt begins 50 million miles away from Mars and extends for more than 100 million miles towards the orbit of Jupiter. There are a few hundred thousand asteroids in this area but they are so spread out that no spacecraft travelling in the Belt has ever had a collision with an asteroid.

Chasing away the dinosaurs

The Near Earth Asteroid Rendezvous (NEAR) space probe had a soft landing on an asteroid on 12 February 2001. NEAR found that an asteroid is a solid object but it has craters like the Moon. An asteroid is nothing but a chunk of rock, and scientists believe that the asteroids in the Asteroid Belt are the material for a planet that never formed. Ceres, discovered in 1801, is the largest asteroid known to mankind; it is so large that it is now called a dwarf planet. Some scientists believe that a large asteroid had hit the surface of Earth once upon a time and caused dinosaurs to disappear.

ISRO's destination

Vesta is a large asteroid. Scientists in ISRO consider Vesta to be the primary target option for study. They are curious to learn how the Sun's family of planets and minor bodies originated. Remote sensing of the asteroids will help them understand the evolution of asteroids and comets, early solar system processes, meteorite-asteroid connection, and physical and chemical properties of asteroid material.

The payload for a mission to Vesta would include remote-sensing instruments for imaging, mineral and chemical mapping of the surface and sub-surface. The mission to Vesta will be a long-duration mission and studies of energetic particles, radiation and fields in interplanetary space will be a part of the mission.

HOMAGE TO THE HEAD OF THE FAMILY

The Sun is the head of our solar family. It is actually a huge fireball. The fire started with a fusion reaction 5 billion years ago. To keep the fire going, the Sun burns 600 million tonnes of hydrogen per second. This makes the Sun shine. We can see all the planets and moons when they reflect sunlight.

Is the Sun dying?

Scientists estimate that the Sun has enough hydrogen to burn for another 5 billion years. After that, the Sun will die and start to cool off. What will happen to human life on Earth then? After all, the food chain on Earth starts with sunlight through photosynthesis. We depend on the fossil fuel created from the energy of the Sun. Earth goes around the Sun at a distance so conducive to human life. It is neither too hot nor too cold. The temperature is just right to retain water as a liquid. These are things essential for the existence of life.

Aditya

ISRO is interested in studying the Sun from close quarters. It plans to send a satellite to the Sun. It will be called Aditya, and will concentrate on studying the Sun's corona.

MANNED MISSION

Lost and found

Today, ISRO's confidence is at an all-time high. In 2007, ISRO proved that it could recover and bring an orbiting space capsule back to Earth. This was the Space-capsule Recovery Experiment (SRE). The SRE capsule, which weighed 550 kg, was sent to orbit in space and after twelve days it was brought back and made to fall in the Bay of

Bengal. When the frogmen of the Indian Coast Guard recovered the capsule from the sea, it was in good shape.

The success of the SRE experiment gave ISRO an understanding of how orbiting capsules could re-enter Earth and how they could be recovered. No other country in the world has successfully recovered an orbiting spacecraft in the first attempt. ISRO has now mastered the re-entry and recovery technology which is a pre-requisite for manned missions.

Sending men on a space cruise

ISRO's plan now is to send manned missions carrying a crew of two astronauts and return them safely to a predefined destination on Earth. It will be launched by GSLV. For the first flight, the mission is planned for some orbits lasting a few hours, but the system is designed in such a manner that the mission could even be extended up to seven days.

At the upper end of the rocket there will be the orbital vehicle and the crew escape system. The orbital vehicle will have two parts. One is the Crew Module and the other, the Service Module. The Crew Module will be the control centre and living quarters of the crew for the entire mission. The Service Module will have all the sub-systems required for orbital operations like the thrusters, solar panel and propulsion system.

The magic craft

Human space flight is an activity which ISRO proposes to pursue for future generations by undertaking more ambitious missions. The plan is to follow up the first flight with long duration missions to the low earth orbit using GSLV-Mk III rocket. GSLV-Mk III is a more powerful rocket. Another module can be added to this rocket—a habitat module. The habitat module will be an orbiting laboratory where the astronauts can perform experiments for long durations.

Indians on the Moon?

In a few years, ISRO hopes to land astronauts on the Moon. With that, an Indian would have set foot on the Moon. Achieving a technical marvel like that would clearly demonstrate India's scientific superiority and strength.

Chandravihar

Building a habitat in space would provide a platform for human survival in space. It would also provide support to carry out fundamental and applied research there. Manned missions are important because they help us in constructing and maintaining such platforms in space. One day, we can go to the Moon, establish a base and live there. Scientists at ISRO have even suggested a name for the habitat: Chandravihar.

Imagine what a beehive of activity the Moon would then be! Engineers would be busy building a permanent colony that will require the best of their skills. Geologists can make field trips and study the Moon. Astronomers can look at the universe from the Moon. Scientists can study the solar wind and look at the Moon's past locked away in its regolith. Biologists can grow plants in lunar greenhouses. Mining and chemical engineers would extract resources from the moon rocks. Our world out there on the Moon would be a very exciting one but, mind you, it would be full of challenges.

Only a transit camp

But, in one sense, this lunar colony would only be a transit base. It would be the first settlement from which fearless souls will go away on trips that will take them to other planets and beyond, deep into the universe. With Chandrayaan 1, India's space odyssey has just begun. In course of time, ISRO would have taken us on long, adventurous journeys deep into the universe.

A JOURNEY THROUGH TIME

Our universe is 13.7 billion years old. And it is mighty big. At least as big as the distance that light can travel in 13.7 billion years. A light year is the distance that light can travel in one year, and that distance is roughly 9.46 trillion km. What it means is that the deeper we look into space, the further we will be looking back in time. One day, when ISRO takes us to the very edge of our universe, we will see the beginning of time. What a fantastic journey in time and space that will be!